TRENDS IN MATERIALS SCIENCE RESEARCH

TRENDS IN MATERIALS SCIENCE RESEARCH

B.M. CARUTA

EDITOR

Nova Science Publishers, Inc.
New York

NOTICE TO THE READER

The Publisher has taken reasonable care in the preparation of this book, but makes no expressed or implied warranty of any kind and assumes no responsibility for any errors or omissions. No liability is assumed for incidental or consequential damages in connection with or arising out of information contained in this book. The Publisher shall not be liable for any special, consequential, or exemplary damages resulting, in whole or in part, from the readers' use of, or reliance upon, this material.

This publication is designed to provide accurate and authoritative information with regard to the subject matter covered herein. It is sold with the clear understanding that the Publisher is not engaged in rendering legal or any other professional services. If legal or any other expert assistance is required, the services of a competent person should be sought. FROM A DECLARATION OF PARTICIPANTS JOINTLY ADOPTED BY A COMMITTEE OF THE AMERICAN BAR ASSOCIATION AND A COMMITTEE OF PUBLISHERS.

LIBRARY OF CONGRESS CATALOGING-IN-PUBLICATION DATA
Trends in materials science research / B.M. Caruta (editor).
   p. cm.
Includes bibliographical references and index.
ISBN 1-59454-367-4
1. Materials science. I. Caruta, B. M.
TA403.T743
620.1'1--dc22
2005
2005011404

Published by Nova Science Publishers, Inc. ✦ New York

CONTENTS

Preface   **vii**

Chapter 1   Cooperative Anisotropic Theory of Ferromagnetic Hysteresis   **1**
               Carl S. Schneider

Chapter 2   Image Reconstruction and Geometrical Analysis of Three-   **49**
               Dimensional Fracture Surfaces in Materials
               Manabu Tanaka, Yosuke Kimura, Ryuichi Kato, Junnosuke
               Taguchi and Naohide Oyama

Chapter 3   Optical Properties of Pd-Free Au-Pt-Based High Noble Alloys   **81**
               Takanobu Shiraishi

Chapter 4   Yield Surface of Shape Memory Alloys   **101**
               W.M. Huang and X.Y. Gao

Chapter 5   Recent Advances in Photoluminescence of Amorphous   **117**
               Condensed Matter
               Jai Singh

Chapter 6   The Thermoluminescent (TL) Properties of the Perovskite-   **147**
               Like $KMgF_3$ Activated by Various Dopants: A Review
               C. Furetta and C. Sanipoli

Chapter 7   The Equivalent Simple Cubic System: A New Tool to Model   **157**
               Real Powder Systems under Compression
               J.M. Montes, F.G. Cuevas, J. Cintas, J.A. Rodriguez and E.J.
               Herrera

Chapter 8   Barium Ion Leaching and its Effect on Aqueous Barium   **191**
               Titanate Tape Properties
               Dang-Hyok Yoon and Burtrand I. Lee

Chapter 9   Surface Processes in Charge Decay of Electrets   **215**
               G. A. Mekishev

Chapter 10   Theoretical Studies of Q1D Organic Conductors: A Personal   **241**
                Review
                Vladan Čelebonović

Index   **257**

PREFACE

Materials science includes those parts of chemistry and physics that deal with the properties of materials.

It encompasses four classes of materials, the study of each of which may be considered a separate field: metals; ceramics; polymers and composites. Materials science is often referred to as materials science and engineering because it has many applications. Industrial applications of materials science include processing techniques (casting, rolling, welding, ion implantation, crystal growth, thin-film deposition, sintering, glassblowing, etc.), analytical techniques (electron microscopy, x-ray diffraction, calorimetry, nuclear microscopy (HEFIB) etc.), materials design, and cost/benefit tradeoffs in industrial production of materials. This book presents new research directions in a very new field which happens to be an old field as well.

Measurement of B(H) curves of mild steel, nickel, cobalt, Terfenol and Othonol have lead to an analytic macromagnetic model in precise agreement and deeper understanding of ferromagnetic hysteresis in terms of saturation magnetization, coercive field, reversible susceptibility and domain cooperation. Sample design, experimental control and analysis are described including details of homogeneous field, stress and magnetization, shape demagnetization, eddy currents, coercive smearing, magnetoelastic vibrations and strain gage use and interpretation for both magnetostriction and stress measurement. The theory is observed in Chapter 1, from virgin and reversal hysteresis curves and reversible susceptibility. Approximations are made to describe minor loops which further reduce to the Rayleigh laws. The effects of stress before and after magnetic field change are derived from magnetoelastic energy and coherent action of Brown's micromagnetic stress field and Bozorth's global magnetostrictive stress field, with excellent agreement with experiments. Verified predictions include stress variation of coercivity, remanence and maximum susceptibility as well as a stress crossover on the saturate hysteresis curve due to domain structure. Brown's and Bozorth's stress fields are found to act in all stress processes and are found to be nonlinear for large stress and magnetization due to their effect on domain structure. Bozorth's stress field is predicted and observed to cause creeping of stress loops and the stress demagnetization effect up to remanence where the Villari reversal occurs. Vector fields and stresses have been applied and vector magnetizations measured to extend this model to three dimensions. Rate dependent eddy current and temperature effects are described theoretically using effective fields and material anisotropy effects. The change in elastic modulus of ferromagnetic transducers can be derived from this theory so that robotic actuators can be smoothly controlled and nonlinear transducers accurately interpreted. The behavior of ferroelectric transducers can also be described by changing the variables of this theory. Application of this model beyond the physics of materials is possible.

Stereo matching method is known as a non-contact method for analyzing the geometry of many objects of different scales ranging from macroscopic geography of earth to sub micron surfaces of materials. The principle of the method is based on the extraction of three-dimensional information using a couple of two-dimensional images, namely, a stereo pair. Computer-aided stereo matching method has recently been developed for the reconstruction of various kinds of three-dimensional images. Kimura et al. have recently developed a new stereo matching method based on the coarse-to-fine format for the three-dimensional image reconstruction with a reasonable accuracy in a short processing time. In Chapter 2, three-dimensional images were reconstructed by the new stereo matching method on several kinds of fracture surfaces such as fatigue fracture surfaces in metals and impact fracture surfaces in ceramics.

Fractal geometry created by Mandelbrot has been applied to the quantitative description of self-similarity in complex figures of objects by the fractal dimension. Tanaka et al. have developed a program of the box-counting method for the fractal analysis of the three-dimensional fracture surfaces in metals and ceramics. The fractal dimension of the fracture surface was estimated by the box-counting method and was correlated with the specific microstructure in a certain length scale range of the fractal analysis. The results of the three-dimensional fractal analysis were compared with that of the two-dimensional fractal analysis in metals and ceramics. A good correlation was found between the result of the three-dimensional fractal analysis and that of the two-dimensional one.

Finally, a new method of mapping shape parameters such as fractal dimension and surface roughness was proposed for the investigation of fracture mechanisms and fracture processes in materials. This new method can detect characteristic patterns on a fracture surface, which cannot be observed solely with a scanning electron microscope, and may lead to a new research field, namely, three-dimensional fractography. This approach was also applicable to the evaluation of the area proportions of specific fracture surfaces in materials.

The effects of the addition of Pt to Au on its optical properties and the effects of small additions of various base metals to a parent binary Au-10 at.% Pt alloy on its optical properties were investigated in Chapter 3 by means of spectrophotometric colorimetry. Spectral reflectance curves for the mirror-polished flat samples were collected in the wavelengths ranging from 360 to 740 nm. Three-dimensional color coordinates, *i.e.* L^*, a^*, and b^*, in the CIE (Commission Internationale de l'Eclairage) 1976 $L^*a^*b^*$ (CIELAB) color space were obtained to specify the color of the sample.

The strong decolorizing effect of Pt was evidenced when Au was alloyed with Pt. This decolorizing effect was considered to be due to the formation of "virtual bound states". The alloying addition of a small amount of base metals with a high number of valence electrons to the Au-10 at.% Pt alloy effectively gave a gold tinge to the parent Au-10 at.% Pt alloy. Analysis of spectral reflectance curves for all the alloys revealed that the position of the absorption edge in the visible spectrum was not affected by the alloying elements within the limit of experimental error. However, the slope of the spectral reflectance curve at its absorption edge near 515 nm (approximately 2.4 eV) systematically increased with increasing number of valence electrons per atom, e/a, in the alloy. As a result, with increasing e/a-value a chromaticity index b^* (yellow-blue coordinate) markedly increased and a chromaticity index $a*$ (red-green coordinate) slightly increased, giving a gold tinge to the parent Au-Pt alloy. It was found that the number of valence electrons per atom in an alloy plays an important role in determining color of Au-Pt-based high noble alloys. This information is expected to be useful in controlling the color of Pd-free Au-Pt-based high noble alloys.

An analytical and experimental study was conducted to investigate effects of gravitational acceleration on the weld pool shape and the microstructural evolution for Ni, 304 stainless steel, and Al-4 wt.% Cu alloy. Effects of welding heat source were investigated in Chapter 4 by using laser beam welding (LBW) and gas tungsten arc welding (GTAW). As the gravitational level increased from low gravity (LG~1.2 g) to high gravity (HG~1.8 g) by using NASA's KC-135 aircraft, the weld pool shape for 304 stainless steel was affected considerably during GTAW; a 10% decrease of penetration and 10% increase of width. However, insignificant change in the microstructure and solute distribution was observed at gravitational levels between LG and HG for the 304 stainless steel welds. A binary alloy composed of larger density difference solute is recommended as a potential material system to amplify the gravitational effects on the microstructural evolution. For this reason, the GTAW microstructure on Al-4 wt.% Cu alloy was investigated to determine the effect of gravitational orientation on the weld solidification behavior. Gravitational orientation was manipulated by varying the welding direction with respect to gravity vector, i.e., welding upward opposing gravity (ll-U) and downward with gravity (ll-D) on a vertical weld piece and welding perpendicular to gravity ($\perp$) on a horizontal weld piece. The ll-U weld showed 22% larger weld pool area than that of the $\perp$ and ll-D welds. Under the same welding conditions, larger primary dendrite spacing (λ_1) in the ll-U weld was observed near the weld pool surface and the fusion boundary than the case of $\perp$ and ll-D welds. The ll-D weld exhibited different solidification morphology and abnormal 'S' shape of solidification rate during its growth. For GTAW on Ni and 304 stainless steel, significant effects of gravitational orientation were observed on the weld pool shape that was associated with weld surface morphology and convection flow correspondingly. This was accomplished through GTA welding and a numerical study of the welding process. However, the weld pool shape for LBW was mostly constant with respect to the gravitational orientation. Studies of gravity on the welding process are expected to play a significant role in the space-station construction and circumferential pipe welding on the earth.

Chapter 5 investigates three issues about the yield surface (transformation start stress) of shape memory alloys (SMAs). The first is the two-pole phenomenon. We show that this phenomenon can happen in a SMA if there is a variation in the Young's modulus upon the phase transformation. The second is about multiphase transformation in some SMAs. In such a case, the yield surface may be determined by the superposition method. We compare our predictions with the experimental results of two SMAs. The last one is the difference in the yield surface between the phase transformation and martensite reorientation. It reveals that despite that there is not much difference in the yield surface in the (σ_1-σ_2, σ_3=0) plane between the phase transformation and martensite reorientation, the yield surface of the phase transformation is a slant cone if there is a volume variation in the phase transformation, while that of the martensite reorientation is always a slant column as no volume change is associated in the martensite reorientation. Ni-based superalloys are used as turbine blades or vanes in gas turbines or jet engines in airplanes. It is well known that the fcc and $L1_2$ coherent two-phase structure plays important roles in strengthening Ni-based superalloys at high temperatures. However, the melting temperature of Ni-based superalloys is around 1573 K, and it is difficult to improve the temperature capability above 1473 K. To improve the heat efficiency of gas turbines or jet engines, it is necessary to improve the temperature capability. Ir was selected as an alternative material because its melting temperature is 2720 K. However, no systematic research on Ir or Ir-based alloys has been conducted yet. Therefore, we started a basic investigation of the microstructure and mechanical properties of Ir-based alloys. First, we found that the fcc and L12 coherent structure appeared in some binary Ir-based alloys. This suggests that there is a high potential for the use of Ir-based alloys as high-temperature materials that can be used at around 2073 K. The phase relationship, microstructure, and mechanical properties, such as compressive strength and compressive creep properties between 1473 and 2073 K, of binary, ternary, and quaternary Ir-based alloys have been investigated. Chapter 6 is a report that focuses mainly on the phase relationship, microstructure, precipitate morphology, mechanical properties, fracture mode, and fabrication of Ir-based alloys. In addition to Ir-based alloys, other platinum group metals-base alloys are also introduced.

Thermoluminescence (TL) dosimetry has been developed to the stage that it represents a key technique in absorbed dose determination. TL dosimetry has found a very important use in clinical, personal and environmental monitoring of ionizing radiation. Interest in radiation dosimetry by the TL technique has resulted in numerous efforts seeking production of new, high performance TL materials. Of the many materials that have been produced and studied, several are now commonly used as thermoluminescent dosimeters (TLD). The aim of chapter 6 is to present a review concerning the TL dosimetric characteristics as well as the kinetics parameters of the perovskite-like compound $KMgF_3$ in combination with various dopants. $KMgF_3$ has been growth and extensively studied since 1990 in the Physics Department of Rome University "La Sapienza" and the reported thermoluminescence characteristics have shown this phosphor to be a very good candidate for ionizing radiation dosimetry.

A comprehensive review of the recent developments in studying the photoluminescence in amorphous materials is presented in Chapter 7. Experimental results are reviewed and new theoretical results are derived. Four possibilities of radiative recombination for an exciton are considered: (i) both of the excited electron and hole are in their extended states, (ii) electron is in the extended and hole in tail states, (iii) electron is in the tail and hole in extended states and (iv) both in their tail states. Rates of radiative recombination corresponding to each of the four possibilities are derived: a) within two-level approximation, and at (b) non-equilibrium and (c) equilibrium conditions. It is found that the rates derived under the non-equilibrium condition, have no finite peak values with respect to the photoluminescence energy. Considering that the maximum value of the rates derived at equilibrium gives the inverse of the radiative lifetime, the latter is calculated for all the four possibilities in a-Si:H. The theory is general and is expected to be applicable to all amorphous materials, including organic materials used for fabricating light emitting devices.

In Chapter 9, a new way of approaching some aspects related to powder compaction is proposed. This new route consists on modeling a real powders system (particles of unequal size and form) by means of a system of spheres with simple cubic packing. This latter system suffers the same type of deformation that the real system (uniaxial, biaxial or triaxial), and possesses a porosity degree that in some aspects, makes it equivalent to the real one.

The evolution of particles shape, the effective contact area between particles and the effective path to be traveled into the powder aggregate by the electrical or thermal flow are of great interest in order to establish the connection between both systems (real and cubic). These topics are studied for the simple cubic system and the real system.

Initially, the deformation of the simple cubic system of spheres, subjected to uniaxial, biaxial and triaxial compression, is studied in detail. This study is essentially carried out by a geometric method (*the inflationary sphere model*), eluding the stress-strain problem. Equations to determine how the radius of the inflationary sphere increases as the porosity diminishes are deduced. Besides, equations determining the contact areas of a sphere with its neighbors, as a function of the porosity during the deformation, are established. These relationships suggest the introduction of a new concept: the *normalized* or *relative porosity* (the porosity to the initial simple-cubic-packing porosity ratio), that has been revealed extremely useful in subsequent studies.

A parallel study on real powder systems is carried out, focused on two questions: (i) the effective contact area determination (the real contact area that supports the truly applied pressure, or that serves as passage to the electric current, for example), and (ii) the effective path determination (the length of the shortest way, avoiding pores, that connects two points of both bases of the compact, and that, in principle, would be the way to follow by the electrical or thermal flow). In both questions, use of the aforementioned relative porosity is made, now defined as the quotient between the porosity and the tap porosity of the starting powder. The most relevant proposed equations and predicted results are validated through experimental measurements on real powder aggregates and sintered powder compacts, as well as contrasted with results found in the literature.

After the study of the simple cubic and real systems, the equivalence equations to calculate the porosity that the simple cubic system should have to be considered equivalent to the real one, or vice versa, are established.

Finally, some problems, related to the compaction and electrical conduction of powder aggregates, are studied and solved by means of the new described technique. Proposed solutions to these problems are validated by experiences on real powder systems, and contrasted with data reported by other authors.

In Chapter 10 the authors present three processing aspects to meet the challenges in MLCC processing in aqueous media. One is the understanding of Ba^{2+} ion leaching behavior in water which is known to be one of the drawbacks of water-based slip systems. The second is the effect of excess Ba^{2+} ion on tape properties by using an external Ba^{2+} ion source. The third is the potential solution for the reduction of the amount of Ba^{2+} leaching by using a polymeric passivation agent layer (PAL). EDTA titration method was shown to be in determining the amount of Ba^{2+} ion leaching from $BaTiO_3$ in water. The greater extent and the faster rate of Ba^{2+} leaching were found at the lower solution pH. The excess free barium ions expressed by means of the Ba/Ti ratio adversely affected most tape properties. Increase in the slip viscosity, porosity, agglomeration, and along with a decrease in mechanical properties and green/sintered density were found with the increase in the Ba/Ti ratio. An effort was made to correlate these phenomena with Ba^{2+} leaching in water for realistic MLCC applications. To passivate $BaTiO_3$ surface from Ba^{2+} ion leaching, PAL was formed by drying the slurry after adding a commercial polymeric dispersant. Compared to the conventional dispersant adding method, this PAL method was more effective in reducing the amount of Ba^{2+} leaching. Based on these results, we made practical recommendations for aqueous processing of $BaTiO_3$ powders.

Chapter 11 is a review of the investigations devoted to the role of the sample surface in electret charge decay. Obtaining of electrets was carried out in a corona discharge. The set-up comprises trielectrode system – a corona electrode (needle), a grounded plate electrode and a grid (control electrode) in between. Metal mask with numerous apertures was used instead of the grid for obtaining electrets with island surface charge distribution. The electret surface potential was measured by the method of the vibrating electrode with compensation.

We studied the influence of the anisotropy, and pollution of quartz electrets on the surface potential decay. The behaviour of polymer electrets with island and uniform surface charge distributions was also studied. The results show that at low temperatures and high humidity the drift of the charges on the surface prevails over the transport processes through the bulk of the electret.

The results of studying the behaviour of polymer electrets obtained in corona discharge performed in different gas media or placed under condition of various pressures lower than atmospheric are reported. It was shown that the time dependence of the surface potential is described well by the differential equation for desorption and the dependence of the surface potential on pressure is satisfactorily described by an equation that is analogous to the Langmuir law of adsorption.

Next, the boundary surface potential of electrets according to Paschen's law was determined. An equation for the boundary surface potential was given and a "universal" curve was drawn. Experimental results show that if the surface potential was higher than the boundary surface potential, the charge decay was much more rapid.

Finally, the percolation model was used to analyse electret surface discharge of electrets stored at various conditions of controlled relative humidity for 250 days.

All investigations performed show that under certain conditions surface processes should not be neglected and to extrapolate the results under conditions different from those of obtaining them is incorrect.

The aim of this contribution is to review some aspects of theoretical studies of a family of quasi one-dimensional (Q1D) organic conductors known as the *Bechgaard salts*. In order to make it personal,it will retrace the evolution of the author's interest in this field.

The generalized aim was the calculation of the electrical conductivity of the Bechgaard salts and gaining some knowledge on the equation of state and thermal properties of these materials. Chapter 12 ends with some ideas about the possible future developement of the field.

In: Trends in Materials Science Research
Editor: B.M. Caruta, pp. 1-48

ISBN: 1-59454-367-4
© 2006 Nova Science Publishers, Inc.

Chapter 1

COOPERATIVE ANISOTROPIC THEORY OF FERROMAGNETIC HYSTERESIS

Carl S. Schneider[1]

Physics Department, U.S. Naval Academy, Annapolis, MD 21402

Abstract

Measurement of B(H) curves of mild steel, nickel, cobalt, Terfenol and Othonol have lead to an analytic macromagnetic model in precise agreement and deeper understanding of ferromagnetic hysteresis in terms of saturation magnetization, coercive field, reversible susceptibility and domain cooperation. Sample design, experimental control and analysis are described including details of homogeneous field, stress and magnetization, shape demagnetization, eddy currents, coercive smearing, magnetoelastic vibrations and strain gage use and interpretation for both magnetostriction and stress measurement. The theory is observed from virgin and reversal hysteresis curves and reversible susceptibility. Approximations are made to describe minor loops which further reduce to the Rayleigh laws. The effects of stress before and after magnetic field change are derived from magnetoelastic energy and coherent action of Brown's micromagnetic stress field and Bozorth's global magnetostrictive stress field, with excellent agreement with experiments. Verified predictions include stress variation of coercivity, remanence and maximum susceptibility as well as a stress crossover on the saturate hysteresis curve due to domain structure. Brown's and Bozorth's stress fields are found to act in all stress processes and are found to be nonlinear for large stress and magnetization due to their effect on domain structure. Bozorth's stress field is predicted and observed to cause creeping of stress loops and the stress demagnetization effect up to remanence where the Villari reversal occurs. Vector fields and stresses have been applied and vector magnetizations measured to extend this model to three dimensions. Rate dependent eddy current and temperature effects are described theoretically using effective fields and material anisotropy effects. The change in elastic modulus of ferromagnetic transducers can be derived from this theory so that robotic actuators can be smoothly controlled and nonlinear transducers accurately interpreted. The behavior of ferroelectric transducers can also be described by changing the variables of this theory. Application of this model beyond the physics of materials is possible.

---

[1] E-mail address: schneide@usna.edu

Introduction

The history of ferromagnetism may have begun with the discovery of the lodestone whose force on iron bodies was observed to vary with its size and shape. Later observations published by Gilbert [1] about 1620 showed magnetic field, stress and temperature have the strongest effects on ferromagnetism, vividly portrayed in his sketch of a blacksmith beating a hot iron bar aligned with the Earth's magnetic field. Ampere, Faraday, Gauss and Maxwell united the description of electric and magnetic phenomena in the nineteenth century [2], enabling the use of electric current to control and the ballistic method to measure the magnetic state of samples. In the same century Joule, Villari, Kelvin and Ewing [3] quantified the effects of stress on ferromagnets, giving sketches, numbers, tables, plots and algebra in their descriptions. Following the discovery of the electron by Thompson in 1896, Weiss [4] developed his domain theory of ferromagnetism and Curie explained the effect of temperature in disabling the domains above a critical temperature. These domains were seen through Bitter patterns using a microscope a few decades later. In 1935 Preisach [5], [6], [7], [8] expressed hysteresis through a mathematical convolution of elementary magnet states over a distribution function, and described magnetization changes by imposing the effects of anisotropy, conductivity and temperature. The role of domain walls in the change of magnetization was understood through thermodynamics [9] in the 1930's leading to Brown's theory [10] of micromagnetism and magnetoelasticity and enabling detailed computations of domain wall movement within polycrystalline grains, shown in Fig.(1). Stoner and Wohlfarth [11] developed a model of magnetization flipping between crystalline axes in single domains within small grains. Neel and his student Lliboutry [12] observed details of ferromagnetic hysteresis including "couplage" [13] or interaction, "bascule" [14] or cooperative change in susceptibility and "reptation" [15] or creeping. They expanded on Kondorsky's [16] domain interaction explanation of Rayleigh's [17] hysteresis law with the concept of coupling through a mean domain field but continued to mathematically express this interaction through higher powers of applied magnetic field. This tendency of scientists to describe the response of a system in terms of the external agent is due to anthropocentric pride. Since 1986 Jiles [18], [19] has used a model of ferromagnetic hysteresis based upon a paramagnetic gas modified by a coupling field proportional to the deviation of magnetization from the anhysteretic. Jiles describes domain coupling with five parameters fit to observed B(H) curves and predict magnetization changes. Engineers and mathematicians interested more in modeling transformer hysteresis than physical understanding have used variations of the hyperbolic tangent function to fit hysteresis curves [20, 21]. Lack of precision, simplicity and physical reality in these mathematical descriptions has left ferromagnetic hysteresis models unchanged in undergraduate textbooks for a century.

Advance in scientific understanding was the object of Plato's Allegory of the Cave [22] in which perception evolves from shadows to brilliant light. Scientific language is indebted to Descartes [23] who demanded concise mathematical description of observations and insights. We thank Locke and Hume for openness to scientific truth through skeptical empiricism and insight based on cause and effect. We are grateful to Kant's "Critique of Pure Reason" for openness to transcendental insights. Occam's razor requires that causal models be as simple as possible and this has been expanded by followers [24] of Piaget and Prigogine to view leaps of understanding as phase transitions which condense or reduce the degrees of freedom of a model in approaching cognitive equilibrium. Our macromagnetic model [25] of ferromagnetic hysteresis is based upon two concepts: domain cooperation and crystalline anisotropy. Hysteresis curves are normalized to saturation magnetization and coercive field, making our theory applicable to most magnets. Physical simplicity and wide applicability under changes in field, stress and temperature enhance our model's utility and our understanding of ferromagnetism. The style of our model is distinctly anthropomorphic in its concepts and terminology as encouraged by Goethe [26]. This is satisfying to the author and enhances its applicability to social science. The cooperative anisotropic model was conceived through three decades of support by the Office of Naval Research and the Naval Surface Warfare Center. While immersed in independent research on ferromagnetic hysteresis, the concepts and data of other physicists lead to my own experimental and theoretical insights. This research was stimulated by cross-fertilization not only from other sciences and engineering but also from philosophy, long distance running, photography and singing. As in Pirsig's [27] Zen and the Art of Motorcycle Maintenance, I have during insights sensed the magnet's behavior, transcending my measurements and observations. The model thus conceived is simple, physical and analytic, having no fitting parameters. Derivation of this model from Maxwell's equations and thermodynamics starting from atomic electrons is unlikely due to the number of degrees of freedom which must be condensed, but may be possible starting from interacting domains. I submit this work for public review in the hope that it may enhance understanding of ferromagnetism.

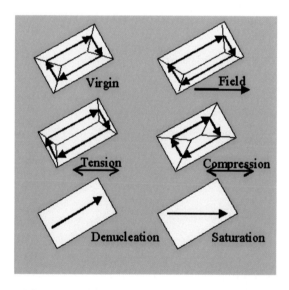

Fig.(1) Domain structure of ferromagnetic grains varies with both applied field and stress. Magnetic field moves 180 degree domain walls causing net magnetization and an external field influencing domain walls in other grains, cooperatively increasing magnetization. Tension parallel to applied field increases 180 degree walls while compression increases 90 degree walls. After wall denucleation magnetization rotates away from preferred cubic axes reducing magnetostriction in steels and causing the Villari reversal of stress effect on magnetization.

In order to speak to as broad an audience as possible, this chapter uses many scientific "languages," including verbal description of physical cause and effect. Numerical data is used to indicate representative values of the quantities in the mksA or SI system. Graphical description through figures includes both sketches and plots of the quantities described in equations. These quantities include both physical and normalized or dimensionless variables, which simplify equations and plots and may better represent physical cause. An example of such normalization from basic physics is the acceleration of a block on a horizontal surface with friction. The free body diagram and Newton's second law informs us that the normal force, N, equals the weight, mg and the friction force, μN, equals the body's mass times its acceleration. Our dimensionless representation of the acceleration is $a/g = \mu$, which displays the physical causes of the acceleration. Similarly, a block sliding on a frictionless incline has dimensionless acceleration $a/g = \sin\theta$ where θ is the angle of incline above horizontal. These dimensionless equations are invariant to transformation of units. In modern physics velocity can be normalized to the speed of light, momentum to mc and energy to mc^2. In thermodynamics [28], the ratio of energy densities is significant and dimensionless variables have long been used. Ferromagnetism is of interest to physicists, who use SI units of Tesla for magnetic induction, B, and A/m for magnetic field, H, and engineers who use emu units of Gauss for B and Oersted for H. We calculate the quantities B and H through Faraday's and Ampere's laws, but the physics of ferromagnetism is due to the applied field, H, and the density of magnetic dipoles, M, which best represents the magnetic state to this author. Magnetization density is maximum when all atomic magnetic dipoles are aligned. The average magnetic moment per atom and density of atoms per volume then give the saturation magnetization, M_s, and the dimensionless ratio of these quantities, M/M_s, no longer depends on the units used to measure it. The characteristic magnetic field in ferromagnetism is that for reversing magnetization, called the coercive field, H_c, and magnetic fields may be represented by the dimensionless ratio, H/H_c. The rate of change of magnetization with magnetic field, or magnetic susceptibility, is dimensionless in SI units but can be normalized to its characteristic value, the initial susceptibility. Stress in magnetoelastic ferromagnets, whose coercive field is due to residual stress and magnetostriction, can be normalized to this residual stress. Finally, magnetostriction, although dimensionless, can be normalized to its characteristic value, saturation magnetostriction. Equations in this work will be given in both physical SI form and causal dimensionless form in the hope of increasing the beauty of physics and understanding of ferromagnetism.

A caution must be offered concerning materials whose coercive field, chemical composition and crystal structure are not homogeneous, or whose stress state and domain structure are not in equilibrium. Hysteresis curves for these materials, such as bent iron wire, would be a convolution similar to that used by Preisach modelers. The Preisach allowance for distributions in my opinion postponed understanding of the fractally cooperative nature of ferromagnetism both in the formation of domains and in their hysteretic interaction. The Preisach convolution was likely motivated by knowledge of the wide range of coercivities observed in magnets, as expressed by the terms "soft" and "hard." The distribution of domain wall pinning energies in a single magnet was assumed to cause the continuous versus first order phase transformation as magnetization is reversed. The Barkhausen [29] effect is evidence that individual domains grow discontinuously as a sample's magnetization changes, but its change over a range of applied magnetic field is as much due to increasing mean domain field and its spatial inhomogeneities as to variations in coercivity or hardness within the sample. Our theory represents observation that differential susceptibility is not symmetric when plotted against magnetization but peaks at negative magnetization. Disagreements with our theory are likely to arise from experimental errors, including shape demagnetization, thermal relaxation, sample bending and magnetoelastic vibrations which introduce stress broadening reducing remanence and susceptibility and delaying maximum susceptibility. Eddy currents and inhomogeneous applied fields also smear and centralize the maximum in susceptibility and past observers have concluded that it occurs at the coercive field and zero magnetization. Careful experiments will verify that maximum susceptibility occurs at magnetization prior to the coercive field.

Experimental Design

Ferromagnetic hysteresis is the irreversible change of magnetization with applied magnetic field, literally M(H), but most scientists and engineers plot B(H), shown in Fig.(2) for a typical mild steel. The choice of B and H are due to the experimental method of measuring magnetic induction, B, using a search coil wrapped around the area through which B passes and turns of wire along the entire length of the magnetized body to create uniform H. Magnetic field H is calculated from Ampere's law from currents in wires and the conducting material and magnetic induction B is calculated from Faraday's law of induced electric fields, E, in Eq. (1) where J is current density and σ is sample conductivity.

$$\vec{\nabla}\times\vec{H} = \vec{J}+\sigma\vec{E}$$
$$\vec{\nabla}\times\vec{E} = -d\vec{B}/dt$$

(1)

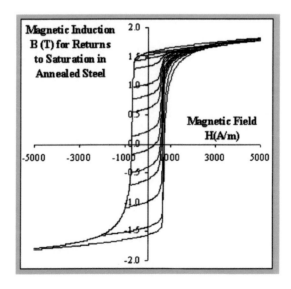

Fig.(2) Magnetic induction of mild steel is shown as a function of applied field including reversals from several fields. Variables B and H are used traditionally due to Faraday's and Ampere's laws.

The differential form of these laws is valid at any point of space while the integral form must be used to determine an equation for specific sample geometry. Integration requires that both B and H be homogeneous or well defined over the region in which they are created and measured else the B(H) curve will be smeared over a range of values and not imply a precise physical relationship. Sample geometries offering nearly homogeneous field are the long cylinder and thin toroid, which can be extended along its axis to become a cylindrical shell, shown in Fig. (3). The cylindrical shell also offers uniform axial tension and compression up to yield stress without buckling and uniform azimuthal stress under internal or external pressure, essential in the study of biaxial stress effects on ferromagnetic hysteresis, which can be called ferromagnetoelasticity. We measure axial B using turns of copper wire wound azimuthally near the center of the cylinder length. Azimuthal B turns may be wound axially along the cylinder length, outside and inside the wall at any point on the circumference although they are usually distributed around the circumference. Amperian H turns must be wound uniformly along the entire length of the sample to create uniform field and flux at the site of the B turns. The cruciform planar geometry used by some researchers [30, 31] in biaxial ferromagnetoelasticity suffers from inhomogeneous field and flux unless the B turns are wound through the sample only near its center, and then the penetrating B holes distort the field and flux uniformity, preventing accurate observation of the laws of physics.

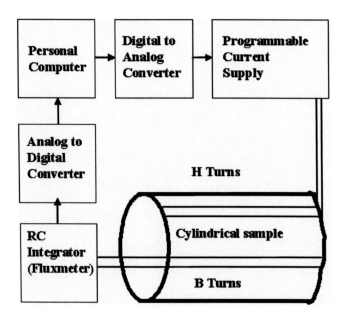

Fig.(3) Cylindrical ferromagnetic samples may be used both axially and azimuthally as shown. A computer not only generates the current waveform but also interprets the input and output voltages to create plots for analysis. Digital integration of Faraday B voltage eliminates the fluxmeter above.

Calculation of the applied axial magnetic field due to windings on a cylinder is possible using Eq. (2) which can be derived from Ampere's law, Eq. (1), for a finite solenoid. The angles are those between the axis and the solenoid ends at the point of observation, n is the H turns per length and I is the current.

$$H = nI(\cos\theta_1 + \cos\theta_2)/2 \qquad (2)$$

Eqn. (2) shows that the applied magnetic field decreases to about half its maximum at the solenoid ends and requires that data be collected at the center where the field is uniform. Unfortunately when a ferromagnetic sample is placed in the solenoid its magnetization goes to zero at its ends, creating magnetic poles and a field which goes from the North end of emerging flux to the South end as shown in Fig. (4). Pole fields are also called the shape demagnetizing fields since they reduce the solenoidal field within the sample. Eq.(3) relates the magnetic induction B to the sum of the magnetic fields from free electron currents, H, and from those bound within the atoms of the sample, M, which is the density of magnetic moments per volume in the sample.

$$B = \mu_0 (H + M) \tag{3}$$

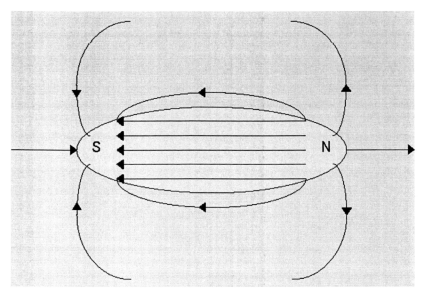

Fig(4). The demagnetizing field of a magnetized ellipsoid is uniform within the body and just outside the middle of its length, where a Hall probe may be placed to sense the net field. The North pole is defined to be the source of magnetic flux.

Taking the divergence of Eq. (3) and using Gauss' law for magnetism (the B field curls around its current sources and is free of divergence) we estimate in Eq. (4) the magnitude of the demagnetizing field at the sample center under the approximation that the magnetization is homogeneous up to the sample end.

$$H_D = -MA / 2\pi r^2 = -\Omega M = -DM \tag{4}$$

The demagnetizing field [32] is a Coulombic field, decreasing as the square of the distance from each pole, and opposes the applied field. The pole strength is MA, the product of the sample magnetization density and its cross-sectional area. The demagnetizing field can be expressed as a demagnetizing factor, D, multiplied by the sample magnetization density, M, and this factor is from Eq. (5) the net solid angle, Ω, that the poles subtend at the center. The shape demagnetizing field has a minimum at the sample center and increases quadratically from there, so B turns must be restricted to the central ten per cent of the length in order for the H field to be homogeneous within one per cent. Accurate values of D factors are given by Bozorth for ellipsoids, for which D is constant throughout the sample. He also plots values of D for cylinders for various sample permeabilities, M/H, since the poles move inward from the ends as permeability increases. The total H field needed to generate a useful B(H) curve exceeds the coercive field of the material, where the saturate hysteresis curve crosses zero magnetization. This field is the sum of the demagnetizing and coercive fields and the crystalline anisotropy field to reversibly approach saturation. In mild steel the field necessary to achieve reversibility is 10 kA/m and 40 kA/m is the anisotropy field in all steels.

These fields can be created using one thousand turns of one millimeter diameter copper wire and a programmable bipolar ten ampere current supply. For our steel samples, saturation induction, B_s = 2.10 T, was measured in our vibrating sample magnetometer and whose saturation magnetization density is M_s =1.67 MA/m. Samples must be long enough that D is near 10^{-3} and the additional demagnetization field is a few kA/m. A hollow cylindrical sample of diameter 25 mm and wall thickness 1.5 mm has an axial flux area πdt near 100 μm^2 and its length must exceed 200 mm in order to overcome hysteresis with field 10kA/m. If this sample is connected to stress fittings of mild steel the magnetic poles may move significantly beyond the sample ends, further reducing the demagnetizing field. But pole motion causes the demagnetizing field to be nonlinear and hysteretic in magnetization. In practice the value of the applied magnetic field, H, is not calculated from Eq. (1) but is measured using an Allegro Hall chip whose output voltage varies linearly over five volts within one per cent for field ranges of plus and minus 300 oersteds or 24 kA/m. Noise of 40 A/m may be reduced considerably through averaging. Relative noise in both magnetic flux and field may be reduced by the square root of the number of points in the average until onset of distortions in the signal. Noise in differential susceptibility, the rate of magnetization to field change, may also be reduced by increasing the width of the change window, again up to the onset of distortion. The net reduction in susceptibility noise thus goes as the three halves power of the number of data points, which can approach one hundred near saturation, and noise is reduced by three orders of magnitude as will be seen in our data.

If the Hall chip is mounted on a shaft it can be oriented within one degree from the axis and susceptibilities up to 10^5 can be measured. When azimuthal flux is stimulated and measured with turns described above, there are no magnetic poles and the magnetic field may be calculated directly from Ampere's law, dividing the ampere turns by the pathlength. The lack of shape demagnetization makes this sample design seem ideal, but in practice the circumference varies over the thickness of the walls and the magnetic field, H, varies by fraction t/r, thickness over radius. Since wall thickness must be at least ten per cent of the diameter to avoid buckling and generate adequate flux signal to far exceed the tens of microvolts thermal noise, the applied field is smeared as is the susceptibility which appears to peak at 10^4 for mild steel. A further advantage of studying axial flux in a cylindrical sample with a Hall chip is that the net field at maximum susceptibility increases one hundredth as fast as the applied field due to the demagnetizing field. The near absence of eddy currents enables measurement of susceptibilities over 10^5. In our research into the character of hysteresis limited maximum susceptibility would weaken our conclusions and the azimuthal or toroidal geometry is used only in the study of biaxial stress effects. One must distinguish between the axial Hall chip noise level and the azimuthal field smearing: the axial field is well defined while the azimuthal field is smeared.

We have found it necessary to prevent degradation of experimental hysteresis curves by magnetoelastic resonance. The LC electromagnetic resonance of the sample and power supply feeds mechanical resonance of the sample through magnetostriction. Strain oscillations induce stresses in the sample which smear the hysteresis curve and reduce and shift maximum susceptibility. Sample inductance, L, and capacitance, C, are given in Eqs. (5) where Al is the sample volume.

$$L = N_B d\Phi / dI = \mu_0 \mu n^2 A\ell$$
$$C = \varepsilon_0 A / r = 2\pi\varepsilon_0 \ell$$

(5)

Self inductance L in the power circuit varies with the permeability, square of the H turn density and sample flux area times length or volume, and is a millihenry times the relative permeability which varies from one to one hundred thousand. Capacitance of the solenoid is ε_0 times the solenoid area divided by its length, and is a few nanofarads. The LC resonance during hysteresis curves thus scans frequencies from ten Hz to a hundred kilohertz. The mechanical resonance of the sample is most easily estimated by the speed of sound in steel divided by twice the length of the sample. This frequency exceeds 20 kHz, the limit of human hearing, for short samples and resonates at a permeability of several thousand. Samples less than four inches long would have ultrasonic vibrations and susceptibility would be smeared without apparent cause. Our solution to this problem when hearing the resonance was to add 100 microfarads across the power supply, which lowered the LC resonance to 100 Hz and the resonant susceptibility to one, the limit of our data range.

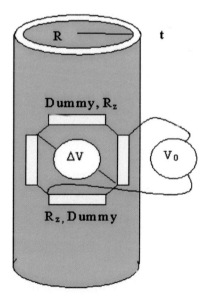

Fig.(5) The Wheatstone bridge used for measuring axial strain on a hollow cylindrical sample is shown. A second bridge is used to measure azimuthal strain. Biaxial magnetostriction and applied stress may be separately determined.

Magnetic flux is measured by connecting the B turns to a LabView DAQ board, digitizing the Faraday voltage and reading it into a data file at up to 5000 data pairs per second. The flux $N_B AB$, some tens of milliwebers, is integrated with an Excel spreadsheet on a laptop computer with an accuracy of one tenth microweber. Uniform drift in the flux reading could be eliminated, by imposing a digital offset to this voltage, within one microweber corresponding to 100 A/m in magnetization density, M, and one in reversible susceptibility. Hysteresis loops are scanned in magnetic field using a digitally generated analog function to control the programmable power supply. A variable frequency sawtooth function is used in order to minimize solenoid heating at large currents below 20 mHz and eddy currents near maximum susceptibility at 1 mHz. The software also reads the Hall chip voltage and output of axial and azimuthal strain bridges, as shown in Fig. (5). Data for each experiment was pasted into an Excel spreadsheet under the first five columns: time, axial strain voltage, Hall voltage, flux and azimuthal strain voltage. In the columns to the right of the data was calculated field, induction, normalized magnetization, susceptibility, stress and other useful properties to best display the meaning of the experiment. In discussing experimental procedure we use an extension of Bozorth's notation for magnetic processes. Demagnetization or equilibration through an increasing frequency and then decreasing amplitude sawtooth field are referred to by D, applied field by H, applied stress by σ, and stress removal by $\bar{\sigma}$. Thus a prestressed equilibrated sample magnetized and then stress removed would be denoted $\sigma D H \bar{\sigma}$ and this could be used to measure the reversible stress field as a function of stress.

Stress was applied through a lever and screw apparatus and measured using axial and azimuthal strain bridges on the cylindrical sample, shown in Fig.(5), each with two active gages and two floating reference gages mounted on an aluminum base to enhance thermal stability. The Wheatstone bridge [33] voltages across one diagonal due to constant ten volts across the other diagonal are given in Eq.(6).

$$\begin{pmatrix} \Delta V_z \\ \Delta V_\phi \end{pmatrix} = \frac{V_0 F_G}{2} \begin{pmatrix} \varepsilon_z \\ \varepsilon_\varphi \end{pmatrix} \tag{6}$$

The bridge factor for a half active bridge is ½ and the gage factor for our 350 ohm gages is 2.12: each microstrain (μm/m) yielded 10.6 μV. Drift in the output due to temperature changes could be removed within a similar accuracy by repeated measurements and magnetostriction could be measured within 0.1 μm/m. Voltmeter noise in the strain bridge readings was several tenths of a microvolt, corresponding to strain precision of several parts in 10^8 when thermal drift was compensated by repeated measurements. For stress measurement the elastic modulus, E = 200 GPa, of our steel was measured using an MTS stress testing machine to within one per cent. Poisson's ratio, 0.30, was checked frequently to give assurance that bending stresses were absent, as they should have been due to the kinematic design of the press, being constrained only in the axial direction by using crossed and centered end pins. Stresses were calculated from strain readings with error less than one mega Pascal using Eq.(7) where ν is Poisson's ratio, measured to be within .01 of .30 for our mild steel.

$$E\begin{pmatrix} \varepsilon_z \\ \varepsilon_\phi \end{pmatrix} = \begin{pmatrix} 1 & -v \\ -v & 1 \end{pmatrix}\begin{pmatrix} \sigma_z \\ \sigma_\phi \end{pmatrix} \tag{7}$$

Combining Eqs. (6) and (7) we solve for the biaxial stress on the sample in Eq.(8).

$$\begin{pmatrix} \sigma_z \\ \sigma_\phi \end{pmatrix} = \frac{2E}{V_0 F_G (1-v^2)}\begin{pmatrix} 1 & v \\ v & 1 \end{pmatrix}\begin{pmatrix} \Delta V_z \\ \Delta V_\phi \end{pmatrix} \tag{8}$$

Typical stress of 100 MPa in our experiments gave a strain of 500 μm/m and a voltage of 5 mV with an error of 2 MPa due to magnetostriction if not corrected. Magnetostriction under load was presumed to be at constant load due to the soft design of our press: displacement of the lever exceeded that necessary to create the strain by roughly a factor of one hundred. Biaxial stresses were imposed upon our hollow cylinders using both internal and external pressure, shown in Fig.(6), as well as axial tension and compression. The stress in the walls of a hollow cylinder of radius R at the surface of either internal or external pressure, P, is given in Eq.(9) in terms of this pressure radius and average radius $\langle R \rangle$.

$$\begin{pmatrix} \sigma_z \\ \sigma_\phi \end{pmatrix} = P\frac{R}{t}\begin{pmatrix} R/2\langle R \rangle \\ 1 \end{pmatrix} \tag{9}$$

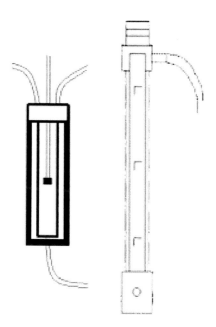

Fig.(6) Hollow cylindrical samples fitted for external (left) and internal (right) pressure as well as axial tension and compression were magnetized axially in the study of the effects of biaxial stress on ferromagnetism. A third sample enabled biaxial field and magnetization measurement under axial stress.

The physical description of ferromagnetic hysteresis is expressed in terms of the magnetization density of the sample and this is calculated from Eq. (3) with a small correction for the gap between the B turn center and the sample flux area. This correction is for the extra magnetic flux sensed by the B turns outside the sample and is best represented by a minimum susceptibility equal to the ratio of the B turn area to the sample flux area. For the azimuthal turn the B area is two or three times the sample area due to wire diameter and openness of the turns. The axial B turns were wrapped around the hollow cylindrical sample and the minimum susceptibility was 13 plus or minus one. Magnetic susceptibility can be measured from one to one hundred thousand and sample magnetization within one per cent using the experimental procedures described above, and this gives confidence in the physical theory that follows.

Cooperative Anisotropic Theory of Ferromagnetic Hysteresis

Rayleigh's [17] description of ferromagnetic hysteresis for minor loops over magnetic fields small compared to the coercive field expressed the cause of the changes as the applied field and the effect as the change in magnetic induction. We rewrite his law in terms of magnetization, field and the Rayleigh hysteresis constant, R, in Eq. (10) so that we can evaluate the differential magnetic susceptibility.

$$M = X_i H + RH^2 / 2 \qquad (10)$$

His observations lead him to conclude that hysteresis is most simply written as a parabolic or nonlinear increase in magnetization with field, and this allows us to write the differential susceptibility as linear in applied field, as shown in Eq. (11).

$$X(H) = \frac{dM}{dH} = X_i + RH \qquad (11)$$

Magnetization changes from the demagnetized state at first linearly with applied field at the initial susceptibility and then more quickly as field increases. Rayleigh approximated the effect of reversals of the direction of magnetic field change by parabolas, shown in Fig.(7), of magnetization with half the curvature or rate of change of susceptibility with field. Rayleigh's hysteresis law was extended in magnetic field values up to the coercive field by Kondorsky [16] in 1942 as expressed in Eq.(12).

$$\Delta M(\Delta H) = 2M(\Delta H / 2)$$
$$X_\Delta(\Delta H) = X(\Delta H / 2) \qquad (12)$$

Both of these hysteresis laws imply that the applied field H is responsible for the nonlinear behavior by making it the independent variable of the magnetization. Rayleigh's initial polynomial expansion was rewritten by Kondorsky in a form that strengthened the field's implied control over ferromagnetic samples' behavior. But the nonlinear behavior of ferromagnetic hysteresis is not due to the applied field but due to its own internal degrees of freedom, most simply represented by the magnetization and domain structure. Our first experimental evidence of susceptibility dependence of a magnet on its magnetic state is plotted in Fig. (8) including the virgin curve and reversals from several magnetic states.

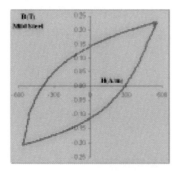
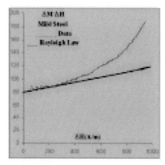

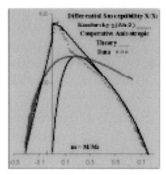

Fig.(7) Hysteresis loop for mild steel at the top and its normal susceptibility in the middle: Rayleigh's linear increase in susceptibility with field and Kondorsky's half field law at the bottom are poor approximations. Eddy current distortion of data is apparent near maximum susceptibility.

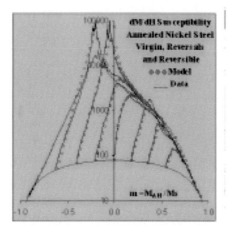

Fig.(8) Differential magnetic susceptibility for the saturate hysteresis loop and several reversals are plotted against the fraction of saturation magnetization showing that susceptibility begins at its reversible value and increases exponentially with magnetization change until saturation dominates. Susceptibility curves increase with cooperation and decrease with anisotropy near saturation.

Differential susceptibility [34] from a reversal begins at the reversible susceptibility which appears to be a single valued function of magnetization. The set of reversible susceptibilities can be written as the initial susceptibility multiplied by a crystalline anisotropy function, χ, of magnetization which has the value one at zero magnetization and decreases monotonically to zero at saturation. In order to describe all ferromagnetic materials, we normalize magnetization to its saturation value as in Eq. (13).

$$m = M_{\Delta H} / M_s$$
$$\chi = X(m) / X_i \tag{13}$$

We measure the sign of the normalized magnetization, m, in the direction of changing magnetic field to render hysteresis equations single valued. Normalized magnetization on a curve of decreasing field starts at negative its measured value and increases toward positive one at saturation, and is identical to increases from negative magnetization. Reversals from either remanence start at positive magnetization and increase to one at saturation. Reversible susceptibility is symmetric in magnetization and crystalline anisotropy functions have been studied [35] by Gans in 1910 and Brown in 1936. They are close to quadratic for isotropic polycrystals, nearly constant for single crystals until saturation, and can be well described by quadratic and quartic terms in magnetization for other materials. All reversible susceptibilities must go to zero at saturation. Measurement of reversible susceptibilities is done by plotting the ratio of magnetization and magnetic field change from any reversal as a function of m as shown in Fig. (9). For changes in m less than ten per cent the curve is linear and its intercept gives the value of reversible susceptibility within one per cent. The set of all values of reversible susceptibility for mild steel is shown in Fig. (10) with a quadratic curve for comparison.

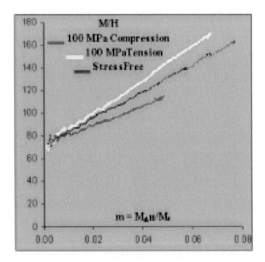

Fig.(9) The ratio of magnetization to applied field on the virgin curve for stresses of zero and 100 MPa tension and compression are linear with small magnetization. Initial susceptibility, due to field induced rotation from crystalline axes, is independent of stress while irreversible susceptibility varies with stress.

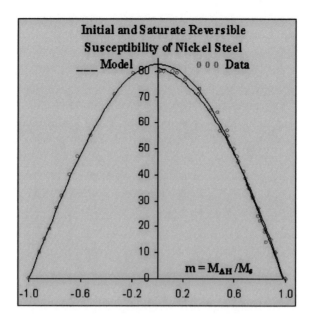

Fig.(10) The set of all values of reversible magnetic susceptibility for isotropic polycrystalline steel is single valued and nearly parabolic in magnetization, having about fifteen per cent quartic term. Observations near zero magnetization on the saturate curve may have thermal viscosity errors.

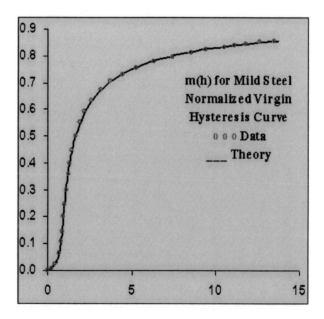

Fig.(11) Normalized virgin magnetization curve, m, versus applied field, h, is precisely predicted by the anisotropic cooperative theory of ferromagnetic hysteresis.

The virgin curve of magnetization from the demagnetized state is shown in detail in Fig. (11). Magnetization begins to increase linearly with applied magnetic field at the initial susceptibility and increases to a maximum slope which then decreases to zero at saturation. Physically, the initial susceptibility is due primarily to reversible rotation of the magnetization away from the polycrystalline axes and toward the applied field. There may also be a contribution from bending of the domain walls shown in Fig. (1). Increase in susceptibility then occurs as the grains are magnetized and their magnetic poles create an external field which acts to increase and decrease the field at surrounding grains. Decreases cause mostly reversible changes while increases push neighboring domain walls past the defects which serve as pinning sites. The cooperative influence of each grain upon its neighbors, represented by the irreversible susceptibility, is thus proportional to the net magnetization and decreases with the threshold or coercive field of the material [36]. The effect of the cooperative domain field upon its neighbors varies as the square of the anisotropy function so that magnetization becomes reversible before saturation. The virgin magnetization can thus be described by the product of the applied field and the sum of the reversible and irreversible susceptibilities as in Eq.(14).

$$M = H(X_{rev} + M\chi_r^2 / H_c) \tag{14}$$

In order to enhance the generality of Eq.(14) for all ferromagnetic materials, we write it in terms of normalized variables, including magnetization, m. The magnetic field, H, can be normalized to the coercive field and must be measured in the direction of its increase as in Eq.(15) to ensure that susceptibility is positive. The physical basis of the coercive field is not explicit and our equation represents all coercive types of ferromagnets.

$$h = H_{\Delta H} / H_c \tag{15}$$

We have used upper case X for M/H so that normalized variables are all lower case. Eq. (14) for the virgin curve may now be written as in Eq. (16). We have here collected the three normalizing constants of the magnet and expressed them in one constant of the material, shown in Eq. (16), which always multiplies magnetization, m, in normalized equations.

$$\beta m = h(\chi_r + \beta m \chi_r^2)$$
$$\beta = M_s / X_i H_c \tag{16}$$

Substituting values for mild steel, M_s = 1.67 MA/m, X_i = 78 and H_c = 660 A/m, we calculate the hysteresis constant beta to be 32. Hysteresis loops having beta of ten or less are sigmoid and those having beta of one hundred or more are square. For materials of coercive field due to residual stress and initial susceptibility due to crystalline anisotropy, beta equals the ratio of their crystalline anisotropy and magnetoelastic energies. It is informative to solve Eq. (16) for the magnetization as a function of applied field, shown in Eq.(17).

$$\beta m = h\chi_r / (1 - h\chi_r^2) \tag{17}$$

Magnetization on the virgin curve can now be seen as the product of the applied field, the reversible susceptibility and a constrained hyperbolic function which describes the cooperation of domains constrained to return to reversibility and then approach saturation.

Another mathematical form of the virgin magnetization is its inverse susceptibility, or reluctivity, shown in Eq. (18)

$$\chi^{-1} = \frac{h}{\beta m} = \chi_r^{-1} - h\chi_r \tag{18}$$

The applied magnetic field reduces the reversible reluctivity until it approaches saturation and becomes reversible again. It may be that the anhysteretic curve is described by subtracting one from the reversible reluctivity in Eq. (18) to remove initial reluctivity. As magnetization increases the reluctivity increases and becomes infinite at saturation. This relationship between reversible susceptibility and the anhysteretic curve suggests there are similarities between our model and that of Jiles. A final form of the virgin magnetization curve is important in extending it to include the effects of stress, temperature, eddy currents and vector representation. This form follows upon inspecting Eq. (16): the anisotropy function is a function of magnetization, m, and it can be separated from the applied field as in Eq. (19).

$$h = \beta m /(\chi_r + \beta m \chi_r^2) \tag{19}$$

This form separates the magnetic field causing change from the sample response through its magnetization in concert with its cooperation and anisotropy. Eq. (19) represents normal equilibrium in which the field and magnetization are directly related, in contrast to susceptibility following reversals which have differential equilibrium. The linear field on the left is an externally imposed anisotropy which by superposition can be generalized to include fields due to stress, eddy currents and thermal excitation, as described later in this work.

The accuracy of the cooperative anisotropic theory of ferromagnetic hysteresis can first be seen from the normal susceptibility curves of steel, cobalt and nickel, shown in Fig.(12). The theoretical curves shown are based on the measured anisotropy functions of these materials, also shown in Fig. (12). A more demanding test of accuracy is a comparison of differential susceptibilities, which rise and fall as magnetization increases. The basis of our model for susceptibility after reversals is the logarithmic plot of mild steel differential susceptibility, shown in Fig. (8) and the cooperative segment is expressed in Eq. (20).

$$\frac{dM}{dH} = X_{rev}(m_r)e^{\Delta M/2XiHc}$$
$$\frac{\beta dm}{dh} = \chi_r(m_r)e^{\beta \Delta m/2} \tag{20}$$

Differential susceptibility begins at its reversible value at a reversal and, after domain walls are nucleated, increases exponentially until it reaches a maximum observable value near one hundred thousand. It then decreases to saturation along an extrapolation of the virgin curve. Reversals from positive magnetization rejoin the virgin susceptibility curve as they approach saturation. Note that the exponent is half the change in magnetization divided by the coercive field and the initial susceptibility. This is the same hysteresis factor of one half observed by Rayleigh and Kondorsky and argued by Neel from a statistical basis. The differential susceptibility of the virgin curve is given in Eq. (21) and may be derived most simply by taking the derivative of Eq. (19) with respect to m, where the denominator is the normal virgin susceptibility, χ_N.

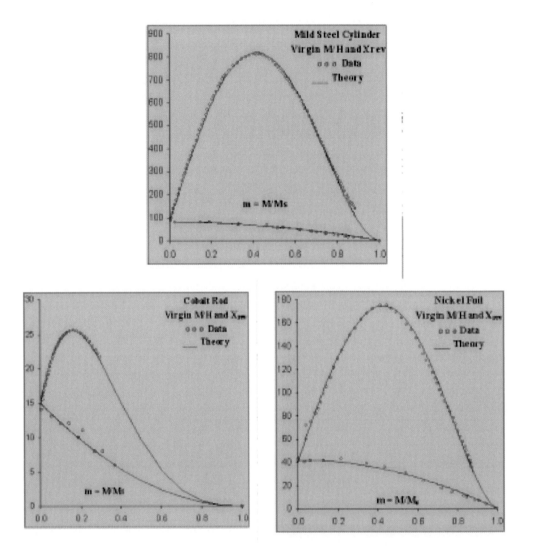

Fig.(12) Normal virgin susceptibility for steel, cobalt and nickel are well predicted by the cooperative anisotropic theory of ferromagnetic hysteresis from the reversible susceptibility and coercive field.

$$\frac{\beta dm}{dh} = \chi_N /(1 - m\frac{d \ln \chi_N}{dm})$$

$$\chi_N = \frac{\beta m}{h} = \chi_r(1 + \beta m \chi_r)$$

(21)

Note that the normal susceptibility increases from its reversible value to a maximum roughly beta times larger and decreases again near saturation to its reversible value. Measured and theoretical differential virgin susceptibility are plotted against the normalized magnetization in Fig. (13) to show the remarkable accuracy of this theory over several orders of magnitude from the demagnetized state to saturation. The differential susceptibility in Eq. (21) can be transformed to that for reversals by changing m to Δm, the change in magnetization from the reversal. Note also that the differential susceptibility is predicted by Eq. (21) to rejoin the reversible susceptibility at m = 0.83 as observed experimentally for cubic crystals. The two stages of hysteresis, cooperative exponential increase in susceptibility of Eq. (20) and anisotropic decrease in susceptibility to saturation in Eq. (21), are piecewise continuous as though each dominates its region. Such mathematics is reminiscent of the wiping out property of minor loop susceptibility discontinuity, the Stoner-Wohlfarth bifurcation and the buckling problem. From reversals fields are integrated from magnetization changes divided by the piecewise continuous cooperative and anisotropic susceptibilities, as in Figs.(14) and (15). Note in Fig.(8) that the slope of the logarithm of the differential susceptibility is constant as magnetization changes but is less for reversals at large magnetization. The constant beta in the cooperative exponent seems modified not by the continuous anisotropy function but by its value at reversal or extremum. This implies that ferromagnetic domain hysteresis from saturation begins with reversible rotation; after domain walls are nucleated near m = -0.83 in cubic crystals cooperation and exponential increase in susceptibility occurs at a rate set by the anisotropy function at this m value. When magnetization becomes positive in the direction of magnetic field change susceptibility switches to its anisotropic function extrapolated from Eq.(22) and approaches saturation. Reversals at large positive magnetization such as remanence show cooperative growth of susceptibility until magnetization saturates with the anisotropy function.

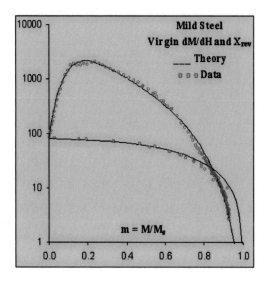

Fig.(13) Differential susceptibility for mild steel is precisely predicted by the cooperative anisotropic theory, even beyond remanence where reversible susceptibility decreases with m faster than quadratic.

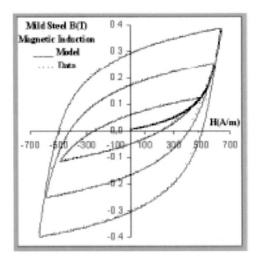

Fig.(14) Integration of cooperative exponential susceptibility accurately predicts hysteresis loops up to coercive fields.

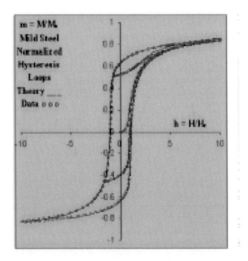

Fig(15) Our theory of ferromagnetic hysteresis precisely predicts the virgin and saturate curves and all reversals and minor loops. Fields are integrated from magnetization changes divided by the piecewise continuous cooperative and anisotropic susceptibilities.

The cooperative anisotropic theory of ferromagnetic hysteresis must agree with Rayleigh's laws for small fields. We will develop two approximations, one for fields up to the coercive field and small magnetization, and one for small fields and magnetizations, the Rayleigh region. For both these coercive and Rayleigh loops magnetization is small enough that the anisotropy function, especially for polycrystals, is essentially one. The coercive region may thus be called the isotropic or cooperative region. The virgin curve in Eq.(20) may then be written as in Eq. (24), which is clearly valid only up to the coercive field at $h = 1$.

$$\beta m = \frac{h}{1-h} \qquad (22)$$

In the Rayleigh region fields are small relative to the coercive field and the virgin curve is written as in Eq. (23) in both normalized and physical variables.

$$\beta m = h + h^2$$
$$M = X_i H + X_i H^2 / H_c \qquad (23)$$

The Rayleigh hysteresis constant is thus $R = X_i/H_c$ in aligning our theory with Rayleigh's. Rayleigh's law of hysteresis lacks both anisotropic and correct cooperative aspects of ferromagnetism. Reversals in the isotropic coercive region obey a differential susceptibility derived from Eq. (20) and shown in Eq. (24).

$$\frac{\beta dm}{dh} = e^{\beta \Delta m / 2} \qquad (24)$$

Eq.(24) can be integrated giving a closed form for m(h) in Eq.(25). These approximations are plotted with experimental data in Fig. (14) showing excellent agreement.

$$\Delta h = 2(1 - e^{-\beta \Delta m / 2})$$
$$\beta \Delta m = -2 \ln(1 - \Delta h / 2) \qquad (25)$$

Finally, from Eq.(25) reversal loops can be approximated by Eq. (26), which agree with Rayleigh's laws in Eq. (10).

$$\beta \Delta m = \Delta h + \Delta h^2 / 2$$
$$\Delta M = X_i \Delta H + X_i \Delta H^2 / 2H_c \qquad (26)$$

The Jiles model of approach to the anhysteretic curve is a computational model with good description of the physical processes of magnetization. It is not precise and does not correctly predict magnetoelastic effects after several reversals in the B(H) curve due to ignoring domain structure and hysteresis. The Preisach model of ferromagnetic hysteresis is a mathematical model based upon integrating an elementary switching response function over a distribution of threshold fields, which is often Gaussian. The wiping out property after two field reversals returns domain wall pinning sites to the same cooperative state and susceptibility as on the original curve and this rule must be part of any hysteresis model, including ours. The congruency property of independence of minor loops upon state of magnetization is in opposition to our anisotropy function and is relaxed with the moving model. The Preisach theorem of identical losses on the ascending and descending branches of a loop is in contradiction with the phenomena of creeping, which is predicted in our theory for magnetic fields by the anisotropy function and for stress by the domain structure effect which follows.

Stress Modulation of Ferromagnetism

In 1981 Neel [37] described the effects of field, stress and temperature on magnetization as some of the unresolved problems in physics. Joule's observation [38] of magnetostriction, λ, under applied field, H, in 1842 and Mateucci [39] and Villari's [40] observation of stress, σ, modulation of magnetic induction, B, was followed by Maxwell's thermodynamic identity [41] in Eq. (27) which followed by taking partial derivatives of energy density, u, with respect to H and σ for a scalar system.

$$u = \lambda\sigma - BH$$

$$\left[\frac{\partial\lambda}{\partial H}\right]_\sigma = \left[\frac{\partial B}{\partial\sigma}\right]_H \tag{27}$$

Equation (27) tells us that the effect of stress in modulating magnetic induction under constant field equals the rate of change of magnetostriction due to magnetic field under constant stress. There is hysteresis in magnetostriction in Eq.(27) which we extract using Eq.(3), the chain rule in Eq.(28) and the data in Fig.(16).

$$\frac{\partial\lambda}{\partial M}\frac{\partial M}{\partial H} = \mu_0\frac{\partial M}{\partial H}\frac{\partial H}{\partial\sigma}$$

$$\frac{dH_\lambda}{d\sigma} = \frac{d\lambda(\sigma)}{\mu_0 M_s dm} \tag{28}$$

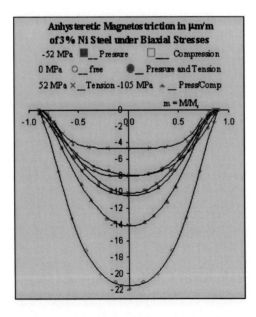

Fig.(16) Magnetostriction in parts per million for mild steel is shown as a function of normalized magnetization for several values of tension and compression as well as for a disc rotated ninety degrees in magnetic fields. These curves are plotted with common remanence and domain structure.

Normalizing magnetic field to the coercive field, stress to residual stress, σ_{res}, and magnetostriction to saturation magnetostriction, Eq. (28) becomes dimensionless Eq. (29) for magnetoelastic ferromagnets with coercive field shown.

$$\frac{dh_\lambda}{d\sigma} = \frac{2d\lambda}{3dm} \qquad\qquad H_c = \frac{3\lambda_s \sigma_{res}}{2\mu_0 M_s} \qquad\qquad (29)$$

For our mild steel saturation magnetostriction from the demagnetized state is 8 μm/m and residual stress is 110 MPa from the effect of stress on remanence given by Bozorth [42] in Eq. (30). The 650 A/m coercive field agrees with Eq. (29) within ten per cent.

$$dB_{rem}/d\sigma = B_s/4\sigma_{res}$$
$$dm_{rem}/d\sigma = 1/4 \qquad\qquad (30)$$

Eq. (30) was confirmed by observed variation of remanence with stress shown in Fig. (17). The magnetostrictive stress field, h_λ, defined in Eq.(29) is the stress integral of the magnetization derivative of magnetostriction, shown as a single valued function of magnetization in Fig.(18) and of stress in Fig.(19). The decrease of dλ/dm to zero under tension, due to the increase of stress insensitive 180 degree domain walls, causes the Bozorth stress field to reach a plateau in tension and a limiting slope with compression as 90 degree domain walls saturate. The effect of domain structure in Eq. (29) is not explicit but is incorporated within magnetostriction and we call this stress field global. It is responsible for the reversible modulation of magnetization by small stresses studied by Bozorth. We are indebted to Birss [43] for his succinct summary of three effects of stress: pressure, anisotropy and domain structure.

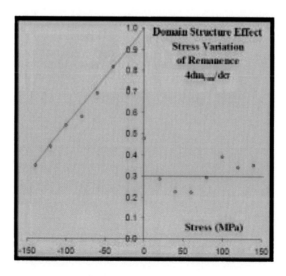

Fig.(17) Residual stress can be determined from the maximum variation with normalized remanent magnetization at small compressive stress in mild steel, which suffers a discontinuous reduction into tensile stress as do Bozorth fields in stress loops.

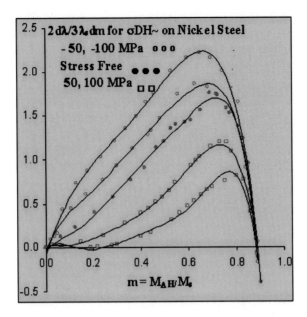

Fig,(18) The magnetization derivative of magnetostriction normalized to 3/2 its saturation value is plotted as a function of magnetization for several stresses. The stress integral of these functions is Bozorth's global stress field. The Villari reversal occurs near remanence. The common slope suggests a return after stress effects to common domain structure.

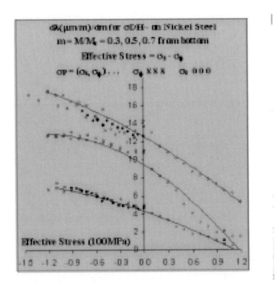

Fig.(19) Stress and magnetization dependence of magnetization derivative of magnetostriction for mild steel are not separable but decrease to zero in tension and increase to 3/2 its unstressed value in compression. Stress orthogonal to field has reversed sign.

Hysteresis in stress modulation of magnetic induction enters through the differential magnetic susceptibility in Eq.(28), described by the cooperative anisotropic theory and this enables description of stress loops. For large stress there is not only cooperative hysteresis but also change in domain structure evidenced by the magnetostriction derivative shown in

Fig.(18). Domain structure change causes the stress field to be nonlinear and causes significant creeping of cycled stress loops. From Fig.(16) the magnetostriction is symmetric in magnetization and its derivative, the global Bozorth stress field, is antisymmetric in magnetization. For magnetization less than the remanent value, when the Villari crystal structure effect begins, the Bozorth stress field approximation in Eq.(31) is linear in magnetization just as a demagnetizing field from the poles at the sample ends. For small stress the slope of the normalized magnetostriction derivative is about 2.2, equal to the magnetostriction of a rotated disc sample relative to that at remanence in Fig. (16), when domain walls are denucleated, and this is roughly λ_{100} or about 2.2 when scaled to λ_s.

$$\frac{dH_\lambda}{d\sigma} \approx 3\lambda_{100}Mf_\sigma / 2\mu_0 M_s^2$$

$$\frac{dh_\lambda}{d\sigma} \approx \lambda_{100}mf_\sigma$$

$$(31)$$

From Eq. (31), reversible tension increases and compression decreases magnetization prior to the Villari reversal in steel, modulated by a domain structure factor, f_σ, of stress which decreases monotonically from a maximum under compression to zero under tension. The first analytic application of our cooperative anisotropic theory of magnetoelasticity is Bozorth's reversible effect of small stresses on magnetization, the $DH\sigma_-$ effect. From the reversible susceptibility and Bozorth stress field in Eq.(31), we find the stress modulation of magnetization in Eq. (32).

$$\beta\frac{dm}{dh_\lambda} = \chi_r$$

$$\beta\frac{dm}{d\sigma} = \lambda_{100}f_\sigma m\chi_r$$

$$(32)$$

Since $\chi_r = 1\text{-}m^2$ for isotropic polycrystals, and $f_\sigma = 1$ for small stresses our theory gives the same expression derived by Bozorth. We have been able to identify the physical basis of the expression as the magnetostrictive Bozorth stress field, linear in m, and the reversible susceptibility, $1\text{-}m^2$. These expressions are both valid up to remanence after which the stress field falls to a negative value giving the Villari reversal following Eq.(29). We show the measured the effect of small stresses on magnetization in Fig.(20) for tension and compression and stress free. Bozorth' use of small stresses precluded his observing that the rate of change of magnetization with stress is discontinuously less for tension than for compression for small magnetization. Stress seems to create an equilibrium domain structure as it changes magnetization when fields are not large enough to control domain structure as at remanence in Fig.(17).

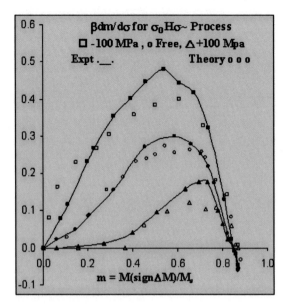

Fig.(20) The rate of change of magnetization with stress is discontinuously less for tension than for compression for small magnetization.

The linear increase of the Bozorth stress field with magnetization for small stresses makes it equivalent to a shape demagnetization field and we define with Spano et al [44] a stress demagnetization factor, D_σ, in Eq.(33) which we modify by domain structure factor f_σ. We normalize D_σ by multiplying by initial susceptibility X_i, and denote it by lower case d_σ or alternatively $1/\chi_\sigma$, the stress change in reluctivity, to avoid confusion with symbol, d, for differential.

$$D_\sigma = \frac{dH_\lambda}{dM} \approx \frac{3\lambda_{100}f_\sigma\sigma}{2\mu_0 M_s^2}$$

$$d_\sigma = \frac{dh_\lambda}{\beta dm} \approx \frac{\lambda_{100}f_\sigma\sigma}{\beta} = 1/\chi_\sigma \tag{33}$$

Net material reluctivity is that from crystalline anisotropy, reduced by cooperation from magnetization change due to applied field plus that from stress. Thus the reluctivity observed for the applied field is reduced by positive stress and susceptibility increases with tension as shown in Fig(20). Similarly compression reduces susceptibility and this was measured for mild steel as shown in Fig.(21) and in more standard B(H) format in Fig.(22). The concept of stress demagnetization requires stress field to be linear in both magnetization and stress, which is a poor approximation in light of Fig.(20). Reluctivity varies greatly due to domain structure changes and we will develop a better representation of the effects of stress.

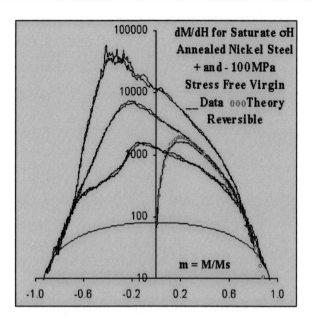

Fig.(21) Near residual stress, 100 MPa tension increases and compression decreases the observed differential susceptibility of mild steel on the saturate hysteresis curve away from saturation. Note the equality of ascending and descending susceptibilities.

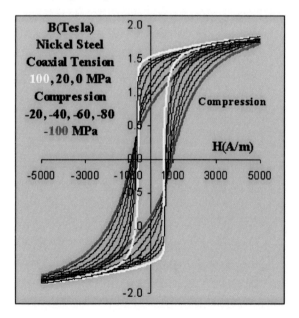

Fig.(22) The slope of the B(H) curves decreases with compression and increases with tension for mild steel. The coercive field increases with compression and decreases with tension and the reverse for remanent magnetization. Note the common crossover.

Large stress dependence of magnetostriction derivative in Fig.(19) requires that we modify small stress Eq. (31) and (32) to include domain structure changes. From Fig.(19) stress and magnetization dependence are not separable and slopes for field dominated domain structure are parallel, as in Eq.(34) which is valid up to remanence and the Villari reversal.

$$\frac{dh_\lambda}{d\sigma} \approx \lambda_{100}(m - \sigma/2)_0^{3/2} \qquad (34)$$

The domain structure factor in brackets is positive definite and less than 3/2 when magnetization is perpendicular to applied field due to compression. The concept of stress controlled domain structure is supported by the data in Fig.(23) where magnetostriction parallel and perpendicular to magnetization are shown. Kashiwaya [45] has described the instability of domain structure depending upon the relative values of stress and magnetic field, in agreement with Fig. (17): analytic expressions such as Eq. (34) are at best approximate. Without stress, perpendicular magnetostriction is half that parallel, described by a magnetostrictive Poisson's ratio ½ as for rubber or isotropic fluids. Poisson's ratio for magnetostriction increases with compression from ½ to 1 as for an incompressible fluid in three and two dimensions as magnetization turns away from the stress axis.

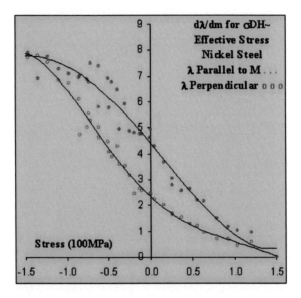

Fig.(23) Poisson's ratio for magnetostriction under compression increases with stress from ½ to 1 as for an incompressible fluid in three and two dimensions, respectively.

The effects of 400 MPa tension and compression on mild steel are shown in Fig.(24) starting from induced and remanent magnetization. Initial effects include the Brown irreversible stress effect which will be studied next. We first study nonlinear stress loops which follow the first stress reversal. Eq. (32) is modified in Eq.(35) to include cooperative or irreversible susceptibility which applies even to the "reversible" Bozorth stress field. It is only for small stresses that the Bozorth effect is reversible. The "irreversible" Brown stress field is ever present but manifested to the extent that positive and negative 90 degree domain walls have different field history, which is not the case for stress loops. Stress fields for normalized tension and compression larger than normalized field are observed to change discontinuously in stress loops and this is represented in Eq.(35) through a separable domain stress factor from Fig.(19) and the crossover in Fig. (22).

$$\left[\beta \frac{dm}{d\sigma}\right]^{H\sigma\sim} = \lambda_{100} m f_\sigma \chi_r e^{\beta \Delta m/2} \qquad (35)$$

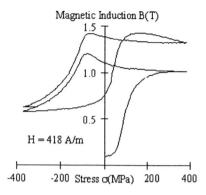

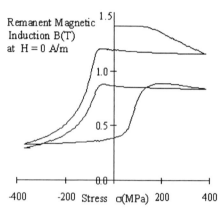

Fig.(24) Large stresses applied to virgin mild steel after magnetic field H and at remanence give irreversible Brown changes in magnetization. Stress cycling gives creeping Bozorth stress loops displaying B(H) hysteresis modified by domain structure changes.

Eq.(35) is valid to remanence and the Villari reversal if the cooperative exponential form of susceptibility changes to Eq.(21) approaching saturation magnetization. Eq.(35) expresses magnetostrictive Bozorth field, domain structure effects and exponential cooperation and crystalline anisotropy prior to saturation. Domain structure factor f_σ used in Eq.(35) and apparent in Fig.(19) can be approximated as in Eq.(36) by a modified Fermi-Dirac [46] function of normalized stress in order to compute the measured stress loops shown in Fig.(24) and (25), which has not yet been done.

$$f_\sigma = 3/(e^\sigma + 2) \qquad (36)$$

Large stress loops require further modification of Eq.(35) due to the wiping out property of hysteresis. When the stress field acting on negative domain walls exceeds twice the applied magnetic field, H, this magnetization component rejoins the reflected or negative virgin curve and magnetization change is given by Eq.(37).

$$\Delta M = (M(H_\sigma + H) - M(H_\sigma - H))$$

$$\Delta M \approx 2HX(\langle H_\sigma \rangle) \tag{37}$$

Eq.(37) expresses differential susceptibility as a function of local domain wall stress field and it is not a function of net magnetization. We are fortunate in having Eq.(17) as a form of virgin susceptibility in terms of field and anisotropy, and we assume it to be a function of magnetization until further studies suggest otherwise. Eq.(17) thus predicts that stress modulated magnetization rises to a maximum and then falls as the stress field exceeds the coercive field and maximum susceptibility, as in Fig.(24). Local susceptibility during release of large stress is cooperative and exponential as in Eq.(20) until it reaches its peak and then is given by Eq.(21), the differential extension of virgin susceptibility. The plateau in large tension still occurs as 180 degree wall domain structure precludes any effect of stress. Rules for hysteresis in many stress and field processes have been tabulated by Brugel and Rimet [47] in the Rayleigh region and these may be translated into reversible and irreversible susceptibility using our theory to extend their discussion to large stress and field.

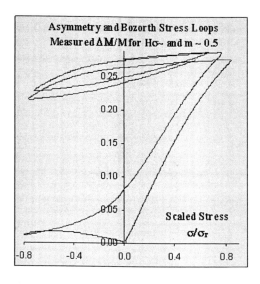

Fig,(25) The stress effect when applied after magnetic field on the virgin curve at more than small magnetization shows Bozorth asymmetry in tension and compression first for irreversible Brown increases and then for creeping stress loops.

Experimental observation of prestress, σH, by Ewing [3] and of post stress, Hσ, by Brown [48] and of small stress cycling, Hσ~, by Villari and Bozorth and of large stress cycling effects by Lliboutry [12] reveal significant differences in both magnitude and reversibility, as in Figs.(24) and (25). The application of stress to a magnet after magnetic field is applied from the demagnetized state gives significantly more change in magnetization than predicted by the Bozorth field. In order to correctly explain irreversible effects of stress, Brown studied magnetoelastic pressure changes across 90 and 180 degree domain walls within grains, finding an energy density change across domain walls within each polycrystalline grain. In Fig.(1) there is one 180 degree wall in each grain separating magnetization aligned in opposite directions. There are four ninety degree walls in each cubic crystalline grain shown, those with magnetization changing direction by 90 degrees. The application of tension moves 90 degree walls to increase 180 degree walls, and this increases magnetization to the right as shown. The two 90 degree walls on top of each grain are called positive walls if the magnetic field is to the right, and the bottom walls are called negative. Fig. (1) is consistent with the energy densities within magnetic domains given in Eq.(38) in physical variables.

$$u = \frac{3}{2}\lambda_{100}\sigma \sin^2 \theta_{\sigma m} - \mu_0 H M_s \cos\theta_{Hm} \tag{38}$$

The quadratic dependence of magnetoelastic energy, the first term, on magnetization direction relative to stress causes there to be no pressure due to stress on 180 degree walls. Further, in the demagnetized state domain structure is symmetric and there is no net magnetization change in a grain under stress. Prestress, however, induces magnetic poles on the ends of larger domains which cause local external fields, grain interaction and a shift in initial susceptibility from prestress. The second term in Eq.(38) is the magnetostatic energy density from which we can solve for the stress field on 90 degree walls by minimizing energy as in Eq.(39).

$$H_\mu = \frac{3\lambda_{100}\sigma}{\mu_0 M_s}\langle \Delta\cos\theta_{\sigma m}\rangle \tag{39}$$

At small magnetization a magnet is isotropic and the difference in domain wall cosines averaged over solid angle hemispheres is positive and negative one half. The Brown micromagnetic stress field of Eq.(39) averages to zero in a demagnetized magnet or after decreasing field reversals leaving history and domain structure symmetric within each grain. After a unidirectional change in magnetic field positive domains experience a cooperatively enhanced susceptibility and negative domains experience reversible susceptibility. The Brown stress field acts with irreversible susceptibility on the positive fraction of domain walls, f_{90}, which was estimated by comparing Eqns.(14) and (16) in Brown's 1949 paper [48] on cubic crystal structured steel to be $1/(\sqrt{2}+2) = 0.293$. This fraction varies symmetrically with magnetization as domain walls are either nucleated or denucleated and is constant near zero magnetization, in contrast to the antisymmetric Bozorth field.

In order to describe both Bozorth's magnetoelastic and Brown's micromagnetic stress fields as well as effects on magnets of magnetic field, temperature, conductivity and sample shape we sum over these imposed anisotropies [49, 50, 51, 52] in the form of thermodynamically defined fields forming a unified field theory of ferromagnetic hysteresis expressed in Eq.(40).

$$\beta dm = \sum_{i=1}^{n}\chi dh_i \tag{40}$$
$$h = h_A + h_\mu + h_\lambda + h_\omega + h_T + h_D$$

The ability to superimpose different fields in ferromagnetism derives from the strictly linear dependence of their effect on magnets. Material susceptibility for these fields are functions of their magnetization, not applied field. Change in magnetization is the integral of the sum of changes over fields due to current, stress, frequency, temperature and shape multiplied by the appropriate susceptibilities for these fields. Total susceptibility applies for H, DM, Bozorth, eddy current fields and Brown for each domain wall type while irreversible susceptibility applies for the dynamic thermal field which causes viscosity. Static thermal effects modify magnetostriction and crystalline anisotropy. Susceptibilities are given by our cooperative anisotropic theory. The magnetic pole field is derived from Gauss' law, stress and thermal fields from thermodynamics and eddy current field from Faraday's law. The work of Brugel and Rimet [47] makes clear the role of hysteresis and stress fields in a large variety of stress and field processes. The significance of their work is not impaired by their restriction to the Rayleigh region since it can be extrapolated using our hysteresis model.

From Eq.(39) we normalize the Brown stress field in Eq.(41) and add domain structure factor f_σ bounded by values zero in tension and 3/2 in compression. We use a differential formulation in anticipation that the Brown field reaches a plateau in tension rather than decreasing to zero as extrapolated from Brown's original definition, but this has not been resolved experimentally. The Brown and Bozorth stress fields are both nonlinear due to domain structure changes with stress.

$$\left[\frac{dh_\mu}{d\sigma}\right]^{H\sigma} = \lambda_{100} f_{90} f_\sigma \tag{41}$$

Magnetization dependence of the Brown stress field, shown in Fig. (30), is due to nucleation and denucleation of ninety degree domain walls and expressed in the fraction f_{90}. At small stress and magnetization the slope in Eq. (41) equals 0.64, as in Fig.(28). The stress field for 110 MPa of compression should be this fraction of 650 A/m, or 420 A/m, about double that observed in Fig. (30) possibly due to nonequilibrium domain structure at the coercive field. When cooperative asymmetry is established by applying field H after demagnetization, D, the Brown stress field in a DHσ process moves fraction f_{90} of positive domain walls further up the virgin curve in Eq.(18) while the negative walls experience reversed net field due to stress and decrease magnetization reversibly. The net effect is that the initial stress field in Eq.(41) acts on the irreversible susceptibility giving magnetization changes described in Eq.(42)

$$\left[\frac{dm}{d\sigma}\right]^{H\sigma}_{Brown} = \lambda_{100} m f_{90} f_\sigma \chi_r^2 \tag{42}$$

Irreversible Brown magnetization increases due to his micromagnetic stress field are linear with net magnetization and the square of the crystalline anisotropy function due to the irreversible susceptibility and act on positive 90 degree domain walls. In the classical Brown experiment shown in Fig. (26), magnetization and stress are small and the anisotropy function

χ_r and stress factor f_σ are nearly one. We predict magnetization increases are self proportionate and the rate of fractional magnetization change with stress is given in Eq.(43).

$$\left[\frac{\Delta m}{m\sigma}\right]_{m,\sigma\ll1}^{H\sigma} = \lambda_{100}f_{90} \tag{43}$$

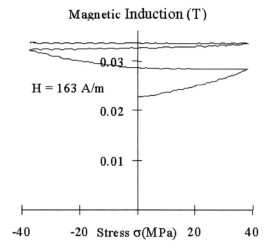

Fig.(26) The classical Brown irreversible stress effect is linear for small tension with horizontal flyback upon stress release. We extend his work here to large stress and to compression to show cooperative nonlinearity in stress and contributions from positive and negative 90 degree domain walls.

This slope is constant and equal to 0.64, the product of 2.2 and 0.293 in excellent agreement with all processes in Fig. (27). Furthermore, as magnetization increases for modest stresses the slope is predicted to decrease with anisotropy as in Eq.(42) for tension whose Bozorth field is negligible. Decreased irreversible stress effect in compression, called the tension/compression asymmetry, is due to coherent Brown and Bozorth stress fields as expressed exactly in Eq. (44). Theory and data are compared in Figs.(28) and (32) in excellent agreement, signaling major success for the cooperative anisotropic theory of ferromagnetism.

$$\left[\frac{\Delta m}{m}\right]_{Irreversible}^{H\sigma} = \chi_r^2(h_\mu + h_\lambda) \tag{44}$$

The ratio of irreversible Brown changes to reversible Bozorth changes of magnetization, seen in Fig. (27), can be calculated by comparing Eqs.(35) and (42). Assuming that the sample has the same domain structure factor, f_σ, and that anisotropy χ_r is near one at modest magnetization, the ratio is βf_{90}, or about 10 to one for mild steel. This constant ratio occurs because the Bozorth field and the irreversible susceptibility are both linear in magnetization. From Fig. (27) reduction in slope at a stress extremum such as the horizontal flyback in release of tension is greater than ten due to changes in f_σ. The domain structure factor is essentially zero when large tension causes pure 180 degree domain walls unresponsive to the Brown stress field. Continuing stress modulation by the Bozorth field involves all domain walls in magnetization change as observed by Finbow and Havard [53], but to an extent described by domain structure f_σ. This increase in stress modulation of magnetization or slope was called "bascule" in French by Neel or "tilting" but was not explained through changes in domain structure. Neel also named the creeping or "reptation" of repeated stress cycles which we compute in the next section using the domain structure factor, f_σ.

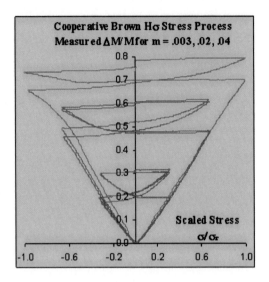

Fig.(27) The Brown micromagnetic stress field on 90 degree domain walls acts with the irreversible susceptibility, which is linear in magnetization from the cooperative anisotropic theory of ferromagnetic hysteresis, causing changes linear in magnetization.

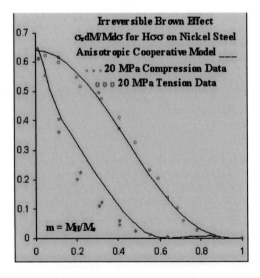

Fig.(28) The tension compression asymmetry, cooperative linearity in magnetization and decrease with magnetization to the Villari reversal of the Brown irreversible change in magnetization is predicted for small stresses by the crystalline anisotropy function.

Stress Fields and Eddy Currents

We have completed above our description of post stress effects and turn to ongoing experimental and theoretical research in order to enable others to understand, undertake and continue this work. The process of prestress and removal, σH and σσH are well described using the concept of stress field and we have measured the total stress field from prestressed B(H) curves in Fig.(22) by taking the difference in the stress free field at magnetization m and that under stress. This procedure is theoretically possible due to Eqns. (19) and (20) in which field enters linearly and all other variables are functions of magnetization and stress but not field. The total stress field is shown in Fig.(29) while its symmetric Brown and antisymmetric Bozorth components are shown in Figs.(30) and (31).

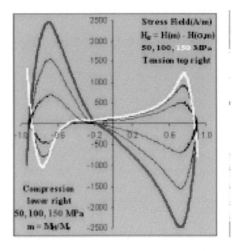

Fig.(29) The difference in magnetic field at identical magnetizations for major hysteresis curves with and without stress gives the total stress field. The fields are the same and near zero at the common crossover.

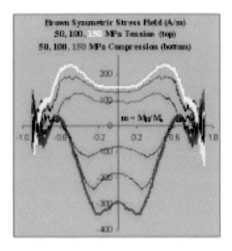

Fig.(30) The symmetric part of the total stress field is the micromagnetic domain field predicted by Brown as responsible for irreversible stress effects. The field goes to zero when domain walls denucleate.

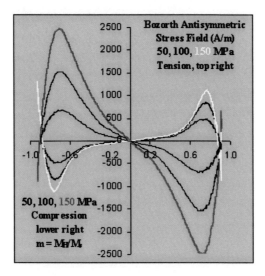

Fig.(31) The antisymmetric magnetostrictive Bozorth stress field is the stress integral of the magnetization derivative of magnetostriction and displays tension compression asymmetry due to domain structure. Note the Villari reversal after denucleation of domain walls.

These plots of physical variables can be normalized to the coercive field, 650 A/m, to show agreement with Eq.(29) for the Bozorth field and (39) for the Brown stress field. Both fields display domain structure changes with stress: tension increases 180 degree walls insensitive to stress while compression increases 90 degree walls responsive to stress. Stress fields in tension reach a plateau while in compression they reach a maximum slope with stress.

At zero magnetization the antisymmetric Bozorth field is zero while the symmetric Brown stress field is not. The Brown stress field is thus the sole stress modulation of coercivity. The rate of change of normalized coercive field is proportionate to the domain structure factor and evidence of domain structure. Remanence occurs at significant magnetization where the Bozorth stress field dominates and domain structure is unstable due to the absence of applied magnetic field as shown in Fig, (17). As shown in Fig.(33), maximum differential susceptibility decreases monotonically with compression in steels and increases with tension until stress fields reach their maximum value and domain structure is dominated by 180 degree walls. At larger stress we have observed a decrease in maximum susceptibility and attribute this not to decreasing magnitude of the stress field but to its direction. Stress stabilizes magnetization along its axis, both parallel and antiparallel, so that susceptibility is reduced. This has not yet been represented mathematically. The common crossover of all B(H) curves under compression shown in Fig. (22) occurs when the net stress field is constant and near zero. This condition is expressed in Eq. (45) by combining Eqs, (31) and (41) for the Bozorth and Brown stress fields assuming that they share the same domain structure factor, f_σ.

$$\frac{dh_\sigma}{d\sigma} \approx f_\sigma \lambda_{100}(f_{90} + m) = 0 \tag{45}$$

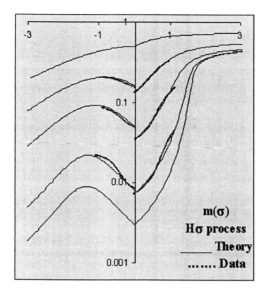

Fig.(32) Effects of large tension and compression on several initial magnetic states are calculated from our theory, which is approximate due to the domain structure factor. Cooperative self proportionate behavior makes a logarithmic plot appropriate.

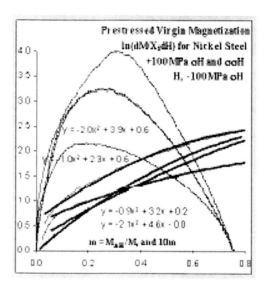

Fig. (33) Prestress causes the initial normalized virgin susceptibility to increase by approximately h_μ, the Brown stress field for small stress. The remainder of the curve is increased by tension and decreased by compression in steel if stress is not released.

We conclude that $m = -f_{90} = -0.3$ under this linear approximation to the Bozorth stress field, in qualitative agreement with Fig. (22). Crossover magnetization as a fraction of saturation is a rough measure of the fraction of stress active ninety degree domain walls. Bozorth concluded in his 1951 book [54] that this saturate crossover was related to that on the virgin curve and the Villari reversal. However, Villari reversals occurnear remanence in iron and steels as magnetization is forced away from the [100] easy crystalline axes and closer to the [111] hard axes which have negative magnetostriction and negative stress field under tension. The saturate stress crossover occurs under compression and is due to domain structure, not crystal structure. We predict that this reversal occurs for other ferromagnetic materials as well as steel where stress sensitive "90 degree walls" are any but 180 degree walls. The Villari reversal occurs near remanent magnetization, after domain walls have been annihilated. It is a reversal not of the Brown stress field, which has reduced to zero, but of the Bozorth magnetostrictive stress field as shown in Figs. (29) and (30).

Studies of the σσH process, application of field to a sample which has been stressed and relieved after demagnetization, appear to have been published only by this author. Most theoreticians and experimentalists must assume that there is no effect since there is no stress, but domain walls have been moved by the prestress field as it changes domain structure and relief of this stress is hysteretic as magnetization changes from cooperative to initially reversible. Review of Fig.(1) reveals that a ferromagnetic grain is a closed flux system only in the demagnetized state. Once stress changes domain structure so that 90 degree domain walls are no longer at 45 degrees to magnetic flux, symmetry is broken and flux escapes the grain to cause intergrain dipole magnetic interactions, or cooperation. Since magnetization is not changing we cannot use Eq.(16) but must use Eq.(17) in terms of applied field, stress field in this case. We can semi-quantitatively estimate that fraction f_{90} of stress active domain walls have experienced negative stress field and an equal fraction positive stress field. Removal of stress brings the two fractions to their corresponding remanent states as shown in Fig.(34). Application of magnetic field h then has the negative fraction starting with cooperation induced increase in susceptibility given in normalized Eq.(46) and in agreement with Fig.(34).

$$[\Delta\chi]^{\sigma H} \approx h_\sigma = f_{90}\lambda_{100}\sigma f_\sigma$$
$$[\Delta\chi]^{\sigma\sigma H} \approx h_\sigma /2 = f_{90}\lambda_{100}\sigma f_\sigma /2$$

(46)

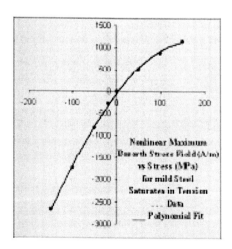

Fig.(34) The Bozorth stress field in mild steel is nonlinear in stress, reaching a maximum in tension as domain structure changes to 180^0 walls insensitive to stress. In compression stress field reaches maximum slope with stress as 90^0 walls dominate structure.

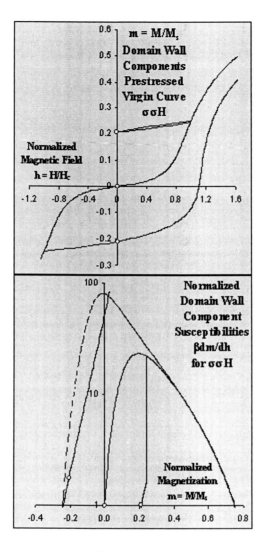

Fig.(35) Domain wall types 90⁻ at m<0, 180⁰ at m=0, and 90⁺ at m>0 follow different magnetization and susceptibility curves due to cooperation between grains in the σσH prestress process resulting in a net increase in initial susceptibility of the virgin curve roughly equal to half the Brown stress field. 90⁻ wall susceptibility rejoins the virgin curve when the Amperian field exceeds the stress field.

When stress is cycled, especially between compression and tension, domain structure changes cause the net magnetization to rise after each cycle much more than that due to field cycling and anisotropy changes. Neel caused this creeping "reptation" in French but did not offer an explanation for it. Qualitatively, in the presence of magnetic field magnetization change for increasing stress begins with larger Bozorth slope than the reversal as in Fig. (38) and grows in proportion to the change in magnetization resulting in greater increase than decrease, creeping upward.

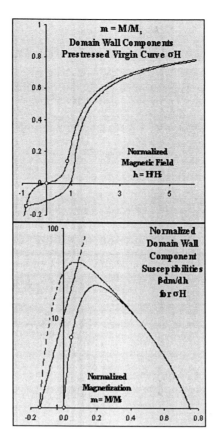

Fig.(36) Prestress causes the three domain wall types in steel, 90^+, 90^- and 180^0, to move so that grains interact cooperatively and net susceptibility of the virgin curve σH increases initially by the Brown stress field and then also by the Bozorth stress field as magnetization increases.

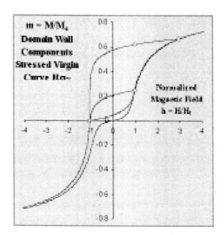

Fig.(37) The Brown irreversible effect of stress is visualized through changes in each of 90^+, 90^- and 180^0 domain wall types. In this example one unit of magnetic field is followed by two units of stress field which moves positive walls to h = 3 and negative walls to -1. Removal of large stress further increases magnetization and subsequent stress cycling has no effect in the pure Brown process.

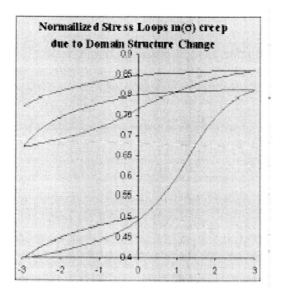

Fig.(38) Creeping of stress loops is predicted through the nonlinearity of stress fields due to changes in domain structure with stress.

The effect of the Brown stress field on magnetization after a field change from any state is to cause the dominant magnetization change in the wall type favored by the field change. Tension or compression applied on the virgin curve always increases magnetization toward the anhysteretic curve until magnetization is sufficiently large that effects are reversible. Then the Bozorth stress field dominates and tension increases and compression decreases magnetization if magnetostriction is positive. At even higher magnetization magnetostriction reverses sign in steels and the Villari effect occurs in which tension decreases magnetization. Applying stress at remanence moves domain walls at higher susceptibility in the direction of the most recent field change so magnetization decreases due to both tension and compression. Application of stress at the coercive field causes rapid decrease in magnetization, again toward the anhysteretic. However, if the applied field has been reduced to zero from the coercive point, the initial effect of stress is to increase magnetization until the stress field exceeds the coercive field and the wiping out rule enables the previous susceptibility. Then magnetization again approaches the anhysteretic curve. This example, reported by Maylin and Squire [55] in 1993, demonstrates that Jiles' law of approach to the anhysteretic is violated and that our theory of cooperative stress fields is correct. Recently Jiles [56] has adopted our two component stress field but neither stress hysteresis effects nor domain wall structure, which weakens his model.

Extension of our theory to three dimensions should follow from the linear dependence on magnetic field in Eq.(14), which can become a vector. Anisotropy is a symmetric function of magnetization and should remain a scalar measured in its own direction. Cooperation or irreversible susceptibility is linear in magnetization and might become a vector. Triaxial stress effects follow from thermodynamic energy density which requires that stress orthogonal to magnetization be negative the product of magnetostriction and stress, where tension is positive, as originally defined by Lord Kelvin [57] using pressurized rifle barrels.

Our cooperative anisotropic theory of ferromagnetic hysteresis has been shown to be precise for the effects of both magnetic field and stress. Through Eq. (19) the theory is linear in applied field and this enables our theory to be extended to a unified field theory including the effects of eddy currents and temperature. Material effects of cooperation and anisotropy are functions of magnetization state rather than applied field, while domain structure may be a function of stress and field, and this is not resolved. Eddy current fields can be computed using both Ampere's and Faraday's laws given in Eq.(1) where σ is the sample conductivity. The eddy current field is due to the electric field induced by the changing magnetic induction, B, due to the net solenoid and induced eddy currents. In our experiment the field of the solenoid was measured directly using a Hall chip so the back emf in the solenoid was automatically included and the incomplete penetration of flux into the sample due to eddy currents was measured through the net flux measured through the secondary or B turns. Ampere's and Faraday's laws can be combined into the diffusion equation describing space and time dependence of the magnetic field, shown in Eq. (47).

$$\nabla^2 \vec{H} = \sigma\mu \frac{d\vec{H}}{dt} \tag{47}$$

When Eq. (47) is applied to axial magnetization in a cylindrical steel sample subjected to an external time dependent magnetic field, the radial variation of eddy current field may be calculated by integrating Eq.(48).

$$\frac{dH}{dr} = \sigma\mu r \frac{dH}{dt} \tag{48}$$

The effects of eddy currents are well known [58] and their mention here is to enable computation of dynamic hysteresis loops using our theory. From Eq. (48) an external applied magnetic field on a conducting ferromagnet causes a magnetic diffusion wave to flow radially inward with no variation around either the minor or major circle of a toroidal ferromagnetic sample. The wave velocity is the coefficient of the time derivative in Eq. (48) and this predicts a diffusion time constant given in Eq. (49), where μ is the maximum permeability over the radius in order to avoid overlapping waves.

$$\tau = \sigma\mu r^2 / 2 \tag{49}$$

$$\Delta B = \mu\Delta H(1 - e^{-t/\tau}) \tag{50}$$

Average magnetization thus penetrates over time as in Eq. (50) due to a step change ΔH in applied magnetic field. From the time derivative of Eq. (50) multiplied by the flux area assuming small field steps and homogeneous permeability, we derive and present in Eq. (51) the voltage induced across N Faraday turns surrounding the flux, first reported by Sixtus and Tonks [59] in 1931 and shown in Fig. (39).

$$V = \pi Nr^2 \frac{dB}{dt} = \frac{2\pi N\Delta H}{\sigma} e^{-t/\tau} \tag{51}$$

$$\mu = \frac{2}{\sigma r^2} \int_{H_i}^{H} e^{-(t-t')/\tau} \frac{dH'}{dH/dt} \tag{52}$$

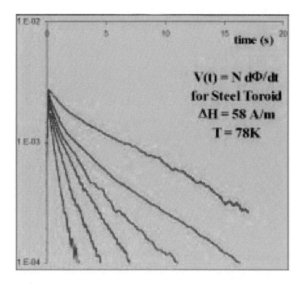

Fig.(39) Faraday voltage due to a constant field step sensed by 100 turns around a steel toroid decreases from a constant predicted by Eq. (51) for 10 MS/m conductivity. Exponential decrease displays a band of time constants varying slightly with permeability.

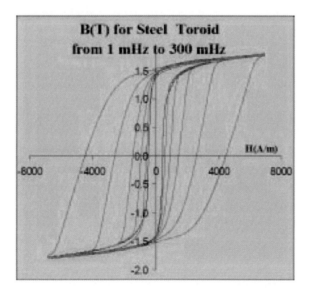

Fig. (40): Ferromagnetic hysteresis curves of a thin steel toroid show increasing coercive field and decreasing permeability for triangular excitation field at rates from 1 mHz for the narrow curve to 300 mHz for the broad curve.

Integrating over differential field, dH, we calculate an effective permeability in Eq. (52) which is delayed in time and appears on a B(H) plot in Fig. (40) to be delayed at the coercive field by the classical eddy current field, $H_\omega = \tau \, dH/dt$. Due to the self proportionate nature of diffusion, with permeability appearing both on the left and right sides of Eq. (52) through the time constant, maximum permeability decreases as the square root of the rate of field change, shown in Eq. (53), and observed coercive field correspondingly increases, as shown in Fig. (41).

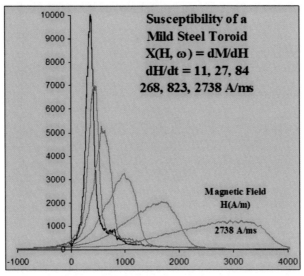

Fig. 41: The susceptibility of dynamic hysteresis curves is the convolution of the quasistatic curve and an exponential delay function of the rate of magnetic field change, whose normalized shape is invariant.

$$\mu_{max}(\omega) = \left[\frac{2\Delta B}{\sigma r^2 dH/dt} \right]^{1/2} \tag{53}$$

Temperature effects on ferromagnetic hysteresis loops are similar to those of stress since each helps to bring the system closer to equilibrium or the anhysteretic curve. Just as stress presses domain walls both parallel and antiparallel to the stress axis resulting in changes in magnetization in the direction of latest field change, temperature causes isotropic pressure on domain walls and is equivalent to an isotropic field whose effects are manifest in the field direction. Thermal relaxation directly reduces the coercive field of the material as does the Brown stress field. Temperature also changes saturation magnetization, crystalline anisotropy and magnetostriction, effects which have been well studied.

The energy dissipated in a ferromagnetic hysteresis cycle can be calculated from our theory but has not yet been done. This heat loss is of interest in magnetic engines and acoustic transducers. The change in elastic modulus of a ferromagnetic transducer under stress, the delta E effect, can be determined from the ratio of stress to strain, shown in Eq.(54).

$$\frac{1}{E} = \frac{d\varepsilon}{d\sigma} = \frac{1}{E_0} + \frac{d\lambda}{d\sigma}$$

$$\frac{d\lambda}{d\sigma} = \frac{d\lambda}{dm}\frac{dm}{dh}\frac{dh}{d\sigma} \tag{54}$$

The net strain from stress is the sum of elastic and magnetoelastic contributions. The second term can be expanded using the chain rule as the product of the $d\lambda/dm$, whose magnetization and stress dependence are given in Figs. (18) and (19), dm/dh, proportional to the susceptibility which is hysteretic and varies with field and stress order, and the stress field $dh/d\sigma$, including both Brown and Bozorth contributions. An initially magnetized sample under tension will increase in size under tension both from the elastic modulus and magnetostriction and will feel softer. The same sample under compression will have less initial decrease in size, feeling stiffer, and then more as compression decreases magnetization and magnetostriction, feeling softer.

Application of the cooperative anisotropic theory to ferroelectric materials follows directly by transforming the variables in Eq.(16) from magnetic to electric quantities as in Eq.(55)

$$P = \varepsilon_0 E(X_{rev} + P\chi_r^2 / \varepsilon_0 E_c) \tag{55}$$

Susceptibility is correspondingly defined by $X = dP/\varepsilon_0 dE$ and the coercive field is that where the polarization reverses on the saturate loop. The validity of this application has been verified using data for several polarization loops. Application to general hysteresis such as capillary condensation [60] of gases in disordered porous solids has been conceived using variables of flow rate and pressure. The flow seems to cooperatively enhance itself giving rise to hysteresis describable using our theory. Social applications of our theory also seem possible. The rise in intensity of an individual speaking at a cocktail party [61] increases nonlinearly with the density of people in a room until a maximum is reached near screaming. As people leave the room and density decreases the intensity of an individual does not decrease due to memory until time relaxes it to normal levels. The productivity of an individual increases nonlinearly with the number of colleagues until functions are fully developed. Decrease in collaboration does not decrease these functions until time or stress returns them to their isolated values. Collaboration between nations enhances individual productivity until it saturates and isolation does not reversibly reduce that productivity. If variables can be identified, our cooperative anisotropic theory of hysteresis can be applied far beyond ferromagnetism.

Conclusion

The cooperative anisotropic theory of ferromagnetic hysteresis has been shown capable of precisely and analytically describing the virgin and saturate curves as well as any minor loop of any material whose coercive field, saturate magnetization and reversible susceptibility function is measured. The effects of stress before magnetic field show as predicted that the symmetric Brown micromagnetic stress field and antisymmetric Bozorth magnetostrictive stress field act coherently to change coercive field, remanence and susceptibility and cause the common stress crossover. Stress applied after magnetic field causes as predicted initial irreversible Brown changes in magnetization which reflect the field history rather than approaching the anhysteretic curve. Magnetization changes are asymmetric in stress due to the Bozorth stress field. For larger stresses two effects occur: first the Brown field acts on both positive and negative domain walls and second domain structure changes, becoming insensitive to tension and more sensitive to compression in steels. When stress is cycled magnetization loops occur and creep as predicted due to nonlinear stress fields. Finally, at magnetization beyond domain wall denucleation the Villari reversal occurs in steels due to negative magnetostriction along the [111] crystalline direction. Further applications of this theory are described and readers are encouraged to pursue them.

Acknowledgements

The author is indebted to the Naval Surface Warfare Center (Carderock), Office of Naval Research, Naval Sea Systems Command and Naval Academy for its support of this research and USNA Physics Department faculty and Martin Sablik of Southwest Research Institute for stimulating discussions. The author is indebted to his mother Viola Schneider for translating the original papers of Mateucci and Villari from French and German. He is indebted to the Brigade of midshipmen for joining him in Socratic dialog in the physics classroom for thirty seven years, to his current Trident Scholar midshipman Stephen Winchell for joining in this search for truth, and to God for revealing the beauty of his universe.

Biographical Sketch

Carl Stanley Schneider was born in Baltimore, Maryland in 1942 and educated throughout northeastern United States and Canada, receiving his B.A. from Johns Hopkins University under advisor Franco Dino Rasetti, codiscoverer of neutron induced fission with Fermi. His Master's degree was on spark chamber measurement of gamma ray energy using π^0 meson decay at the Massachusetts Institute of Technology. His 1968 doctorate under 1994 Nobel Prize winner Clifford Shull was in refraction of thermal neutrons from an iron prism to determine the forward scattering amplitude and S-wave contribution to the magnetic moment of iron. He has been a faculty member of the U.S. Naval Academy Physics Department since 1968. After making several precise measurements of neutron scattering amplitudes at the National Bureau of Standards (now NIST) he began study of stress induced changes in magnetization of steel for the Naval Surface Warfare Center, Office of Naval Research and Naval Sea Systems Command. His three decades of research in magnetoelastcity have lead to this publication. He is the inventor of the closed loop degaussing algorithm, patent number 5,189,590, and recipient of the Frank R. Haig prize of the American Association of Physics Teachers for best paper on the classroom jumping ring, sold by PASCO, and the Charles Rowzee award for best technical paper at the 1996 Mine Warfare Association meeting. He is a member of Phi Beta Kappa, Phi Kappa Phi, American Physical Society, Sigma Xi and honorary life member of the Mine Warfare Society (now SCOT). He is married to the former

Carole Bottom. They have two children, Kathleen and James, and grandsons Stanley and Samuel.

References

[1]   William Gilbert, *De Magnete*; (Basic Books, New York, NY, 1958)

[2]   D.J. Griffiths, *Introduction to Electrodynamics*; (Prentice Hall, Englewood Cliffs, NJ, 1989)

[3]   J.A. Ewing, *Magnetic Induction in Iron and other Metals*; (D. Van Nostrand: New York, NY, 1892)

[4]   C. Kittel and J.K. Galt, *Ferromagnetic Domain Theory*; edited by F. Seitz and D. Turnbull, *Solid State Physics*; (Academic Press: New York, NY, 1956) Vol. 3, pp 437-564

[5]   F. Preisach, *Z.Phys.* **94**, 277 (1935)

[6]   I.D. Mayergoyz, *Mathematical Models of Hysteresis*; Springer: New York, NY, 1991

[7]   E. Della Torre, *Magnetic Hysteresis*; (IEEE Press: New York, NY, 1999)

[8]   G. Bertotti, *Hysteresis in Magnetism*; (Academic Press: New York, NY, 1998)

[9]   R. Becker and W. Doring, *Ferromagnetismus*; (Springer: Berlin, GE, 1939)

[10] W.F. Brown, Jr., *Magnetoelastic Interactions*; (Springer tracts in natural philosophy, Springer, New York, NY, 1966) Vol. 9

[11] E.C. Stoner and E.P. Wohlfarth, *Phil. Trans. R. Soc. London* **A240**, 599-642 (1948)

[12] Louis Lliboutry, Ann. de Phys. **6**, 731-827 (1951)

[13] Louis Neel, Academie des Sciences, Paris, *Comptes-Rendus Hebdomaines des Seances* **246**, 2313-2319 (1958)

[14] Louis Neel, Academie des Sciences, Paris, *Comptes-Rendus Hebdomaines des Seances* **246**, 2963-2969 (1958)

[15] Louis Neel, Academie des Sciences, Paris, *Comptes-Rendus Hebdomaines des Seances* **244**, 2668-2674 (1957)

[16] E. Kondorsky, *J. Phys.* **6**, 93-110 (1942)

[17] Lord Rayleigh, *Philosophical Magazine and Journal of Science* **23**, 225-252 (1887)

[18] D.C. Jiles and D.L. Atherton, *J. Magn. Magn. Mater.* **61**, 48 (1986)

[19] D.C. Jiles, *J. Phys. D* **28**, 1537-1546 (1995)

[20] M. Aribe and J.L. Ille, International J. of Appl. Elec. and Mech. **8**, 273 (1997)

[21] Z. Sigut and T. Zemcik, *J. Magn. Magn. Mats.* **73**, 193-198 (1988)

[22] Plato *The Republic*, Book VII, "Allegory of the Cave", in *Great Dialogs of Plato*, translated by W.H.D. Rouse, edited by P.G. Rouse and E.H. Warmington, (New American Library, Mentor Books, New York, NY, 1956)

[23] Peter Sedgwick, *Descartes to Derrida*; (Blackwell, Malden, MA, 2001)

[24] H. Beilin, Piaget's New Theory and R. Garcia, The Structure of Knowledge and the Knowledge of Structure in Piaget's Theory, Prospects and Possiblities; edited by H. Beilin and P. Pufall, (Lawrence Erlbaum Associates, Hillsdale, NJ, 1992)

[25] Carl S. Schneider, "Anisotropic Cooperative Theory of Coaxial Ferromagnetoelasticity"; *Physica B* **343**, 65-74 (2004)

[26] A.O. Tantillo, *The Will to Create: Goethe's Philosophy of Nature*; (University of Pittsburgh Press, Pittsburgh, PA, 2002)

[27] R.M. Pirsig, Zen and the Art of Motorcycle Maintenance; (Bantam Books, New York, NY, 1984)

[28] H.E. Stanley, Introduction to Phase Transitions and Critical Phenomena; (Oxford University Press, Oxford, UK, 1971)

[29] R.M. Bozorth, Ferromagnetism, (D. Van Nostrand, Princeton, NJ, 1951) p. 524-532

[30] M.J. Sablik, L.A. Riley, G.L. Burkhart, H.Kwan, P.Y. Cannell, K.T. watts and R.A. Langman, *J.Appl. Phys.* **75**, 5673 (1994)

[31] V. Maurel, F. Ossart, Y. Marco and R. Billardon, *J.Appl.Phys.* **93**, 7115 (2003)

[32] Bozorth, op.cit., p 846

[33] C.C. Perry and H.R. Lissner The Strain Gage Primer; (McGraw-Hill, New York, NY, 1962)

[34] Carl S. Schneider, *J. Appl. Phys.* **91**, 7637 (2002)

[35] Bozorth, op.cit., p. 546

[36] C.S. Schneider, *J. Appl. Phys.* **89**, 1281-1286 (2001)

[37] Louis Neel, *IEEE Trans. Magn. MAG*-**17**, 2516-2519 (1981)

[38] J.P. Joule, *Ann. Elec. Magn. Chem.* **8**, 219 (1842)

[39] C. Mateucci, *Ann. De Chim. et Phys.* **53**, 385 (1858)

[40] E. Villari, *Ann. Phys. und Chim.* **126**, 87-122 (1865)

[41] M.J. Sablik, H. Kwun, G.L. Burkhart and D.C. Jiles, *J. Appl. Phys.* **61**, 3799 (1987)

[42] Bozorth, opcit, p. 625

[43] R.R. Birss *IEEE Trans. Magn.* **7**, 113 (1971)

[44] M.L. Spano, K.B. Hathaway and H.T. Savage, *J.Appl.Phys.* **53**, 2667-2669 (1982)

[45] Kenji Kashiwaya, *Jap.J.Appl.Phys.* **30**, 2932-2942 (1991)

[46] C. Kittel *Introduction to Solid State Physics*, 6th edition; (John Wiley, New York, NY, 1986) p. 130

[47] L.Brugel and G.Rimet *J.Physique* **27**, 589 (1966)

[48] Brown, W.F. Jr. *Phys.Rev.* **75**, 147 (1949)

[49] C.S. Schneider and E.A. Semcken, *J. Appl. Phys.* **52**, 2425 (1981)

[50] C.S. Schneider and J.M. Richardson, *J. Appl. Phys.* **53**, 8136 (1982)

[51] C.S. Schneider and Melanie Charlesworth, *J. Appl. Phys.* **57**, 4198 (1985)

[52] C.S. Schneider, P.Y. Cannell and K.T. Watts, *IEEE Trans. Magn.* **28**, 2626 (1992)

[53] D.C. Finbow and D.A. Havard, *Phys.Stat.Sol.* **38**, 541-547 (1970)

[54] Bozorth, op.cit., p. 605-607

[55] M.G. Maylin and P.T. Squire, *J.Appl.Phys.* **73**, 2948-2955 (1993)

[56] D.C. Jiles, *J.Appl.Phys.* **95**, 7058 (2004)

[57] Lord Kelvin, *Mathematical and Physical Papers*, (Cambridge University Press, Cambridge, UK, 1884) p. 370

[58] D.C. Jiles, *J. Appl. Phys.* **76**, 5849 (1994)

[59] K.J. Sixtus and L. Tonks, *Phys. Rev.*, **37** (1931) 930

[60] F. Detcheverry, E. Kierlik, M.L. Rosinberg and G. Taurus, *Physica B* **343**, 303 (2004)

[61] L. Crum, private communication (c. 1974)

In: Trends in Materials Science Research
Editor: B.M. Caruta, pp. 49-79

Chapter 2

IMAGE RECONSTRUCTION AND GEOMETRICAL ANALYSIS OF THREE-DIMENSIONAL FRACTURE SURFACES IN MATERIALS

Manabu Tanaka[*], Yosuke Kimura[**], Ryuichi Kato[***], Junnosuke Taguchi[+] and Naohide Oyama[+]

[*]Research Institute of Materials and Resources, Department of Mechanical Engineering,
Faculty of Engineering and Resource Science, Akita University,
1-1 Tegatagakuen-cho, Akita 010-8502, Japan.
[**]Nippon Software Company Ltd., 31-11 Sakuragaoka-cho, Shibuya-ku,
Tokyo 150-8577, Japan.
[***]Department of Mechanical Engineering, Faculty of Engineering and Resource Science,
Akita University.
[+]Department of Mechanical Engineering, Faculty of Engineering and Resource Science,
Akita University.

Abstract

Stereo matching method is known as a non-contact method for analyzing the geometry of many objects of different scales ranging from macroscopic geography of earth to sub micron surfaces of materials. The principle of the method is based on the extraction of three-dimensional information using a couple of two-dimensional images, namely, a stereo pair. Computer-aided stereo matching method has recently been developed for the reconstruction of various kinds of three-dimensional images. Kimura et al. have recently developed a new stereo matching method based on the coarse-to-fine format for the three-dimensional image reconstruction with a reasonable accuracy in a short processing time. In this study, three-dimensional images were reconstructed by the new stereo matching method on several kinds of fracture surfaces such as fatigue fracture surfaces in metals and impact fracture surfaces in ceramics.

Fractal geometry created by Mandelbrot has been applied to the quantitative description of self-similarity in complex figures of objects by the fractal dimension. Tanaka et al. have developed a program of the box-counting method for the fractal analysis of the three-dimensional fracture surfaces in metals and ceramics. The fractal dimension of the fracture surface was estimated by the box-counting method and was correlated with the specific microstructure in a certain length scale range of the fractal analysis. The results of the three-dimensional fractal analysis were compared with that of the two-dimensional fractal analysis in metals and ceramics. A good correlation was found between the result of the three-dimensional fractal analysis and that of the two-dimensional one.

Finally, a new method of mapping shape parameters such as fractal dimension and surface roughness was proposed for the investigation of fracture mechanisms and fracture processes in materials. This new method can detect characteristic patterns on a fracture surface, which cannot be observed solely with a scanning electron microscope, and may lead to a new research field, namely, three-dimensional fractography. This approach was also applicable to the evaluation of the area proportions of specific fracture surfaces in materials.

Introduction

Mandelbrot et al. [1] first characterized the fracture surfaces of impact-loaded and fractured steels by the two-dimensional fractal dimension (D', $1<D'<2$). Since then, fractal geometry has been applied to the interpretation of physical phenomena including fracture of materials [2-13], grain growth [14] and grain-boundary migration [15] in polycrystalline materials. Microstructural features such as grain-boundary configuration can also be described by the fractal dimension [3, 16, 17], and can be correlated with mechanical properties of materials [3, 16]. The fractal dimension of the fracture surfaces is generally affected by microstructures or fracture mechanisms [2-6]. In metallic materials, the fractal dimension of the fracture surface profile (D_p, $1<D_p<2$) is larger in the ductile fracture surfaces than in the brittle fracture surfaces such as grain-boundary facets in the creep-ruptured specimens [3] or in the fatigued specimens [4]. These fractal dimensions are the global values, which describe a self-similarity in fracture surfaces. Gokhale et al. [7] have reported that there is a local variation in the fracture surface patterns of materials. Difference in the fracture mechanisms may lead to a variety of micro fracture patterns [2,3,8-10]. The fractal dimension of the fracture surface profile (D_p) measured in a certain length scale range decreases with crack growth and with increasing stress in creep 0f an austenite steel [11]. The fractal analysis may give important information about the fracture mechanism or the fracture process in materials.

Characteristic patterns on fracture surfaces of materials can be observed using a scanning electron microscope (SEM) or a scanning probe microscope (SPM) such as a scanning electron microscope (STM) and an atomic force microscope (AFM) [9,12,13]. However, SPMs are not suitable for the observation of complex fracture surfaces, which involve ledges and steps of more than 10 μm in size. On the contrary, computer-aided stereo matching method has been successfully applied to the reconstruction and analysis of three-dimensional images of complex fracture surfaces in materials [18-21]. Kobayashi and Shockey [19] reported that the fracture toughness for crack initiation (J_{IC}) estimated from the reconstructed fracture surface profiles coincided with the value obtained using a conventional fracture test procedure in steels. Stampfl and Kolednik [21] correlated the local fracture toughness estimated from the reconstructed images with the specific microstructures. Their results also indicate that the three-dimensional image reconstruction using the stereo matching method can reproduce the principal features of fracture surfaces and is a useful tool for the estimation of fracture toughness values or for the study of micro fracture mechanisms in materials. Kimura et al. [22] have recently developed a new stereo matching method on the basis of the coarse-to-fine format, which has enabled fast three-dimensional image reconstruction with reasonable accuracy. The authors have applied this method to the estimation of the fractal dimensions of the contours or the fracture surface profiles on the fatigue-fractured specimen of a Cu-Be alloy [23].

It is important to estimate directly the "global" value of the fractal dimension on fracture surfaces (D, 2<D<3) in three-dimensional space, while the three-dimensional fractal dimension (D) can be predicted from the two-dimensional value when a given fracture surface is "isotropic" [24]. Two-dimensional analysis of actual fracture surface profiles can reveal microstructures such as debris, overhang and microcracks linked to the fracture surface, which cannot be detected by three-dimensional fracture surface analysis [11,12,18,19,25]. However, the geometrical information about a given fracture surface in a wide area can be obtained by the three-dimensional fracture surface analysis [18-23]. Combination of three-dimensional and two-dimensional analyses may lead to the further understanding of the geometrical features of fracture surfaces and the fracture mechanism in materials. The authors also applied the stereo matching method to the evaluation of the global fractal dimension of the three-dimensional fracture surfaces in metallic materials and ceramics. Several methods have been proposed for evaluation of the three-dimensional fractal dimension [12, 26-28]. According to the previous study of the two-dimensional analysis [10,11,29,30], the present authors developed a computer program of the box-counting method for the estimation of the fractal dimension of the three-dimensional fracture surfaces [31].

The fractal dimension represents the geometrical complexity of objects regardless of dimension, while the surface roughness estimates the average size of irregularities on the surfaces of objects. Thus, the fractal dimension and the surface roughness have quite different characters in nature. The "local" variations of these shape parameters on the fracture surfaces may give important information on the fracture mechanism, the fracture process and the cause of fracture in materials. The relationship between the fractal dimension and the surface roughness was also investigated for detecting characteristic patterns on the fracture surfaces of materials. Analysis of local fracture patterns on a given fracture surface can be made by mapping shape parameters such as the fractal dimension and the surface roughness as two-dimensional maps and by comparing these maps with each other.

In this study, the fractal dimension was estimated using three-dimensional data of several kinds of fracture surfaces in metallic materials and ceramics obtained by the stereo matching method. Discussion was made on the dependence of the fractal dimension on the microstructures of materials, the analytical method and the size of the analyzed area. The results of the present fractal analysis were also compared with those of the two-dimensional fractal analysis [4, 23, 32]. In this study, computer programs of the fractal dimension map (FDM) by the box-counting method and of the surface roughness map (SRM) were developed for the investigation of the local fracture patterns and the fracture mechanisms of materials. The shape parameter mapping (FDM and SRM) were then applied to the analysis of the reconstructed three-dimensional fracture surfaces of materials. Features and application of FDM and SRM were also discussed on the basis of the analytical results.

Stereo Matching Method

Template Matching

The information in the height direction can be obtained by the stereo matching method using a stereo pair (the basic image and the tilted image) of bitmap images of 256 grey scale levels. The x-coordinate of a point P on the fracture surface, x (on the basic image) moves to that of the corresponding point P', x' (on the tilted image) in the x-direction on the x-y plane (on the base plane) by tilting a specimen around the y-axis (θ), as shown in Fig. 1 [23]. The reference plane of the basic image is the x-y plane and that of the tilted image is the x'-y plane in this figure. The height of the point P from the base plane, h, can be obtained by searching the corresponding point (homologous point) P' in the tilted image. Since the height of these points from each reference plane is the same (Fig. 1), the value of h is given by the following equation with the magnitude of image M:

$$h = \frac{1}{M}\left(\frac{x}{\tan\theta} - \frac{x'}{\sin\theta} \right) \tag{1}$$

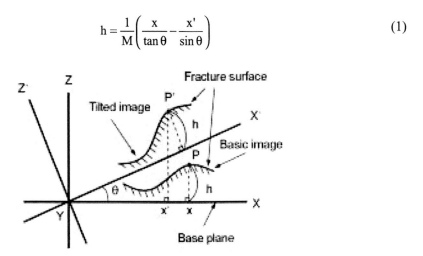

Figure 1 Displacement of the x-coordinate of a point P from x (on the basic image) to that of the corresponding point P', x' (on the tilted image) in the x-direction on the x-y plane (on the base plane) by tilting a specimen around the y-axis (θ) (h is the height of the point P from the reference plane) (From Tanaka et al., [23]. Reprinted with permission of The Iron and Steel Institute of Japan.).

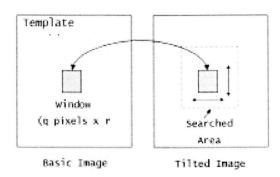

Figure 2 Schematic illustration of the template matching between the basic image and the tilted image (From Tanaka et al., [23]. Reprinted with permission of The Iron and Steel Institute of Japan.).

$$R_C = \frac{\sum_{ij}^{"} (T_{ij} \cdot S_{ij}) \mid N \cdot \overline{T} \cdot \overline{S}}{\sqrt{(\sum_{ij}^{"} T_{ij}^2 \mid N \cdot \overline{T}^2)(\sum_{ij}^{"} S_{ij}^2 \mid N \cdot \overline{S}^2)}} \tag{2}$$

where N is the number of pixels in a template, and T_{ij} and S_{ij} are the brightness (color number) of a pixel located at (i, j) in the template for the window in the basic image and for the imposed area in the tilted image, respectively. The averaged values of the brightness are $\overline{T} = \frac{1}{N} \sum_{ij}^{"} T_{ij}$ and $\overline{S} = \frac{1}{N} \sum_{ij}^{"} S_{ij}$.

Stereo Matching Method Based on Coarse-to-Fine Format

In the template matching, the template size should be large enough for obtaining reasonable results of the matching [18, 29], although the amount of calculation increases with increasing the template (window) size or with increasing the searched area. Therefore, it is necessary to reduce time for template matching, especially on stereo pair with the image size of more than some hundred thousands pixels. In this study, the coarse-to-fine format was employed for facilitating the template matching. Figure 3 shows the procedure of three-dimensional image reconstruction by the coarse-to-fine format [23].

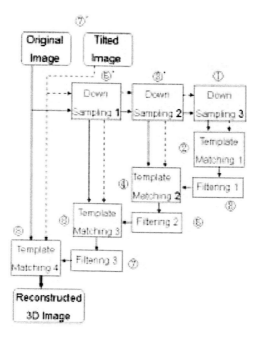

Figure 3 The procedure of the three-dimensional (3D) image reconstruction by the stereo matching method on the basis of the coarse-to-fine format (From Tanaka et al., [23]. Reprinted with permission of The Iron and Steel Institute of Japan.).

The images with the reduced resolution by 1/2, 1/4, and 1/8 were successively produced by "down sampling" (Down Sampling 1 to 3 in the figure) using the original images (the basic image and the tilted image) in this study. The 4-stage template matching (from Template Matching 1 to Template Matching 4) for three-dimensional image reconstruction was then carried out from the image with the lowest resolution obtained after Down Sampling 3 to that with the highest resolution (the original image) in order to detect the corresponding points between the basic image and the tilted image, as indicated by the numbers from □ to □ in the figure. In the matching process, two kinds of the spatial filter, namely, the median value filter and the mean value filter were also used at each step of template matching to eliminate or to reduce the effects of mismatching and noise on the reconstruction of three-dimensional images [22]. The coarse-to-fine format enables fast template matching with a template (window) of small size in small searched area.

Figure 4 shows an example of the stereo pair and the height image generated by the three-dimensional (3D) image reconstruction (stage I fatigue fracture surface of a Cu-Be alloy). The condition of the 3D image reconstruction with the tilting angle of θ=10°(Figs. 4a and 4b) and the template size of 5 x 5 in pixel was chosen in this study, since this condition was found to give a good result of three-dimensional image reconstruction [22]. The relative height on the fracture surface is shown by the brightness (the color number in 256 grey scale levels) in Fig. 4c. Therefore, the relative height increases with increasing the brightness (the color number), and the brightest part in the image indicates the highest part on the fracture surface. A computer program, which was made by Visual Basic version 6.0, was developed and used for the reconstruction, processing and analysis of three-dimensional images in this study. Another computer program used in the previous study was also employed for the fractal analysis of the processed images by the box-counting method [26, 30].

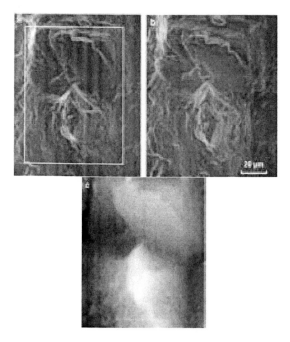

Figure 4 An example of the stereo pair and the height image generated by the three-dimensional (3D) image reconstruction (stage I fatigue fracture surface of a Cu-Be alloy); a is basic image (computed region is enclosed by white lines, 403x467 in pixel); b is tilted image (θ=10°) and c is height image generated by 3D image reconstruction.

Methods of Fractal Analysis

Box-Counting Method

A computer program of the box-counting method was developed for the estimation of the fractal dimension of the three-dimensional fracture surface in this study. The fractal analysis was made using the height data of the reconstructed three-dimensional images [23, 31]. Figure 5 shows the schematic illustration of the box-counting method using boxes of rectangular parallelepiped shape (c is a constant), which cover a three-dimensional fracture surface for the fractal analysis. The number of boxes, N, covering the fracture surface can be related to the box size, r, through the three-dimensional fractal dimension, D_b ($2<D_b<3$), by the following power law relationship [31]:

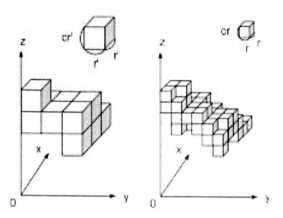

Figure 5 Schematic illustration of the box-counting method using boxes of rectangular parallelepiped shape, which cover a three-dimensional fracture surface for the fractal analysis.

$$N \propto r^{-Db} \tag{3}$$

The fractal dimension, D_b, can be calculated from Equation (3) by the regression analysis using the datum sets of N and r. The height images generated by the stereo matching method were actually used for the estimation of the fractal dimension of the three-dimensional fracture surface in materials [23, 31]. Figure 6 shows examples of the relationship between the number of boxes, N, covering the fracture surface and the box size, r, on materials in the box-counting method (stage I and stage II fatigue fracture surfaces in a Cu-Be alloy and the maximum total strain range in repeated bending ($\Delta\varepsilon_t$) is 0.0171) [34]. The length scale of the fractal analysis is in the range from two pixels to the analyzed area in this case.

Two-dimensional fractal analysis by the box-counting method was also carried out using a computer program on the fracture surface profiles in materials [34]. If N' is the number of boxes involving a profile of fracture surface, crack or a contour and r' is the size of boxes in the two-dimensional box-counting method, there is a relationship similar to Equation (3) between the values of N' and those of r' with the fractal dimension, D_p ($1<D_p<2$), such that $N' \propto r'^{-Dp}$ [24, 26]. Therefore, the length of a profile (or a contour), L, is given by the product of N' and r' [26]:

$$L = N'r' = L_0 r'^{1-Dp} \qquad (4)$$

where L_0 is a constant. The values of L and r' were fitted to Equation (4) to obtain the value of D_p by the regression analysis [4]. Figure 7 shows the examples of the two-dimensional fractal analysis on the fracture surface profiles and the indentation crack profiles [34]. The slope of the linear relationship between L and r' gives the fractal dimension of the fracture surface profiles or the indentation crack profiles, D_p.

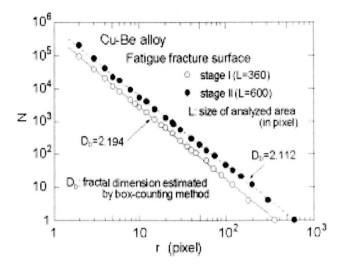

Figure 6 Examples of the relationship between the number of boxes, N, covering the fracture surface and the box size, r, on materials in the box-counting method (stage I and stage II fatigue fracture surfaces in a Cu-Be alloy) (From Tanaka et al., [34]. Reprinted with permission of The Iron and Steel Institute of Japan.).

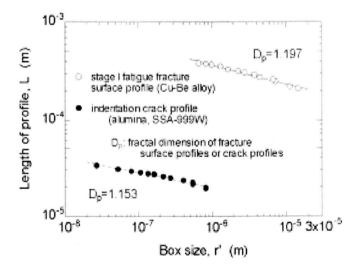

Figure 7 Examples of the two-dimensional fractal analysis on the fracture surface profiles and the indentation crack profiles (From Tanaka et al., [34]. Reprinted with permission of The Iron and Steel Institute of Japan).

Variance Method

Almqvist [12] proposed the variance method for the estimation of the fractal dimension, D_v ($2 < D_v < 3$), of the three-dimensional surface on the images obtained by scanning probe microscopes (SPMs). The fractal dimension of the three-dimensional fracture surface was also estimated by the variance method on some materials. In the variance method, a squared value of the average variance, $\{S(B)\}^2$, is expressed as:

$$\{S(B)\}^2 = \left\langle \frac{1}{B^2-1} \sum_{i=1}^{B^2} (Z_i - \bar{Z})^2 \right\rangle \tag{5}$$

where B is the size of the measurement boxes (pixel), Z_i is the height in each point, $\bar{Z}$ is the average height in the box and $\langle \ \rangle$ denotes averaging over all non-overlapping boxes covering the whole analyzed area. When the distance between data points is the same, there is a relationship between the average variance, $\{S(B)\}$, and the size of the measurement boxes, B, to be expressed with the fractal dimension, D_v, as:

$$\{S(B)\}^2 \propto B^{2(3-D_v)} \tag{6}$$

As shown in Fig. 8, the fractal dimension, D_v, can be calculated from the least square regression line fit to the datum set of $\{S(B)\}$ and B in a log-log plot [34].

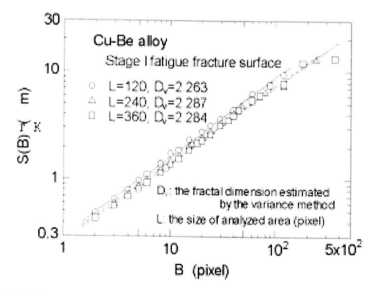

Figure 8 Relationship between the average variance, $\{S(B)\}$, and the size of the measurement boxes, B, in the variance method (stage I fatigue fracture surface of a Cu-Be alloy) (From Tanaka et al., [34]. Reprinted with permission of The Iron and Steel Institute of Japan.).

Mapping of Shape Parameters

The surface roughness, rms, in a given area of m x m in pixel was calculated by the following equation [12]:

$$rms = \sqrt{\sum_{x=1}^{m} \sum_{y=1}^{m} \frac{\{z(x,y)-Z'\}^2}{m^2-1}} \qquad (7)$$

where $z(x, y)$ is the height of a point $P(x, y)$ and $Z' = \sum_{i=1}^{m} \sum_{j=1}^{m} \frac{z(i,j)}{m^2}$. The local values of the shape parameters such as the fractal dimension and the surface roughness can be calculated by using Equations (3) and (7) on a given fracture surface, and can be displayed as two-dimensional maps. Computer programs of the fractal dimension map (FDM) by the box-counting method and the surface roughness map (SRM) were also developed to investigate the micro fracture patterns on fracture surfaces. Figure 1 shows the schematic illustration of the displayed area (k x k in pixel) centered at the calculated region (m x m in pixel) in an image (p x q in pixel) in the calculation and mapping of fractal dimension and surface roughness. The calculated region was moved in both x- and y-directions by k pixels in mapping process. The results of the calculation were finally displayed in color number in the central part of k x k in pixel, and therefore, the brightest area in FDM or SRM corresponds to the part of the largest value of the fractal dimension or the surface roughness. The length scale (r) of the fractal analysis for mapping was in the range from 2 pixels to the size of analyzed area (m pixels). Computer programs were made using Visual Basic 6.0.

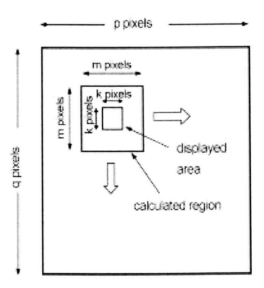

Figure 9 Schematic illustration of the displayed area (k x k in pixel) centered at the calculated region (m x m in pixel) in an image (p x q in pixel) in the calculation and mapping of fractal dimension and surface roughness.

Materials

Metallic materials used for the fractal analysis are the austenitic SUS316 steel (the average grain size is about 1.3×10^{-5} m) [4, 34], a Cu-Be alloy (the average grain size is about 2.4×10^{-5} m) [4] and a pure Zn polycrystals (the average grain size is about 1.2×10^{-5} m) [35]. Ceramics used are a silicon carbide (SiC, the average grain size is about 1.3×10^{-5} m) (Norton NC-430) [31, 32], a commercial alumina (SSA-H) (the average grain size is about 3.0×10^{-6} m) and a commercial mullite (the average grain size is unknown). SSA-H alumina is often used for heat insulator of thermocouples. Fatigue fracture surface of the SUS316 steel and that of a Cu-Be alloy were produced by repeated bending on the rectangular specimen (1.5 mm thickness, 10 mm width and 144 mm length) at the maximum total strain range (on the specimen surface) of 0.0169 for the SUS316 steel [4, 34] and at that of 0.0171 for a Cu-Be alloy, respectively. Creep-rupture experiments of pure Zn polycrystals were carried out under a stress of 14.7 MPa at 373 K [35]. Fracture surfaces of a silicon carbide (Norton NC-430), an alumina (SSA-H) and a mullite were produced by impact loading. Photographs of stereo pairs (the basic image and the tilted image by 10°) on fracture surfaces were taken using a scanning electron microscope (SEM). The photographs were then taken into a computer and were converted to the digital images of 256 grey scale levels. Three-dimensional image reconstruction was carried out to obtain the height data of fracture surfaces using stereo pair images by the stereo matching method.

Results and Discussion

Fracture Surfaces of Materials

Figure 10 shows the scanning electron micrographs of fracture surfaces in metallic materials [34]. These micrographs were used as the basic images for the three-dimensional image reconstruction by the stereo matching method. The direction of fatigue crack growth is approximately from right to left in Figs. 10a, 10b and 10c. Striations and steps are typical patterns of the stage II fatigue fracture surface of the SUS316 steel (Fig. 10a) [4, 36]. In a Cu-Be alloy, the stage I fatigue fracture surface is formed for the most part in a ductile manner with slip steps and dimples except a small amount of grain-boundary facets in the middle part of the micrograph (Fig. 10b), while the stage II fatigue fracture surface is brittle-type with grain-boundary facets and fine dimples (Fig. 10c) [4]. Creep fracture surface of pure Zn polycrystals is composed of equi-axed dimples of various sizes (Fig. 10d) [35]. Figure 11 shows the scanning electron micrographs of fracture surfaces in ceramics [34]. These micrographs of impact fracture surfaces were also used for the three-dimensional image reconstruction. A silicon carbide (Norton NC-430) is composed of many SiC grains of various sizes with a small amount of residual metallic silicon [25, 32]. Brittle fracture may occur along the interface between large a SiC grain and metallic silicon in the silicon carbide (flat region in the middle part of Fig. 11a), while ductile fracture may occur in metallic silicon. Micropores, which may be formed during sintering process, can be observed on the brittle fracture surface of an alumina (SSA-H) (Fig. 11b). A mullite exhibits a complicated fracture surface with micropores (Fig. 11c).

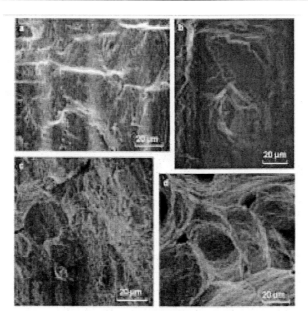

Figure 10 Scanning electron micrographs of fracture surfaces in metallic materials; a is the stage II fatigue fracture surface of SUS316 steel (the maximum total strain range is 0.0169); b is the stage I fatigue fracture surface and c is the stage II fatigue fracture surface of a Cu-Be alloy (the maximum total strain range is 0.0171) and d is a creep fracture surface of pure Zn polycrystals (ruptured at 373 K, 14.7 MPa) (From Tanaka et al., [34]. Reprinted with permission of The Iron and Steel Institute of Japan.).

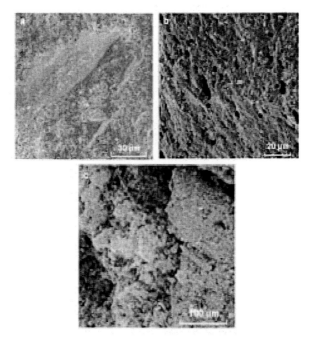

Figure 11 Scanning electron micrographs of impact fracture surfaces in ceramics; a is a silicon carbide (SiC) (Norton NC-430); b is a commercial alumina (SSA-H) and c is a commercial mullite (From Tanaka et al., [34]. Reprinted with permission of The Iron and Steel Institute of Japan.).

Fractal Analysis of Three-Dimensional Fracture Surfaces

Three-dimensional (3D) images of fracture surfaces were reconstructed from the height data obtained by the stereo matching method. Figure 12 shows the bird's eye-view of the stage I fatigue fracture surface in a Cu-Be alloy [34]. This figure corresponds to Fig. 4. The direction of fatigue crack growth is from lower right to upper left in the figure. The fatigue fracture surface of the Cu-Be alloy was formed by ductile fracture mechanism with dimples for the most part, although small areas of the grain-boundary facets were also observed. Thus, the fracture surface is very complicated with ridges, bumps and relatively flat areas. Figure 13 shows the bird's eye-view of the impact fracture surface in a SiC (Norton NC-430) [34]. The impact fracture surface of a SiC has also a complex geometry, since ductile fracture occurred in the retained metallic silicon phase containing fine SiC particles in addition to brittle-type fracture occurred at the interface between a large SiC particle and the retained silicon phase. However, the fracture surface seems to be rather flat compared with that of a Cu-Be alloy (Fig. 12).

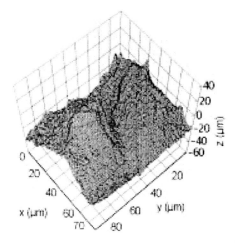

Figure 12 Bird's eye-view of the stage I fatigue fracture surface in a Cu-Be alloy (From Tanaka et al., [34]. Reprinted with permission of The Iron and Steel Institute of Japan.).

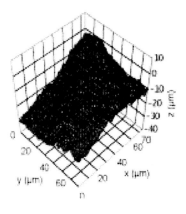

Figure 13 Bird's eye-view of the impact fracture surface in a SiC (Norton NC430) (From Tanaka et al., [34]. Reprinted with permission of The Iron and Steel Institute of Japan.).

The fractal dimension of the three-dimensional fracture surface was at first estimated in the length scale of the fractal analysis from two pixels to the analyzed area. Figure 14 shows the result of the fractal analysis on the three-dimensional fracture surface of a Cu-Be alloy. Some regions for the fractal analysis are also shown in the photograph (numerals are the analyzed areas in pixel) [34]. One pixel corresponds to about 1.81×10^{-7} m in the figure. Both fractal dimensions estimated by the box-counting method, D_b, and by the variance method, D_v, do not largely change with the size of the analyzed area (L in pixel) in this case. The value of D_v is close to 2.30 and is a little larger than that of D_b (about 2.20). Figure 15 shows the result of the fractal analysis on the impact fracture surface of a silicon carbide (Norton NC-430) [34]. One pixel corresponds to about 1.13×10^{-7} m in the photograph (numerals are the analyzed areas in pixel). When the size of the analyzed area is smallest (L=120 pixels), the analyzed area involves only the part of complex geometry (the part of ductile fracture) and shows the highest value of the fractal dimension (both D_b and D_v in Fig. 15).

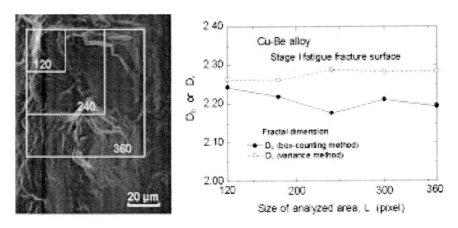

Figure 14 Relationship between the size of the analyzed area (L in pixel) and the fractal dimension of the stage I fatigue fracture surface in a Cu-Be alloy (From Tanaka et al., [34]. Reprinted with permission of The Iron and Steel Institute of Japan.).

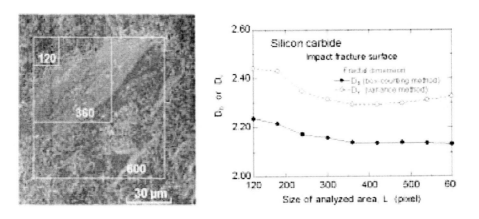

Figure 15 Relationship between the size of the analyzed area (L in pixel) and the fractal dimension of the impact fracture surface in a silicon carbide (Norton NC-430) (From Tanaka et al., [34]. Reprinted with permission of The Iron and Steel Institute of Japan.).

Both values of D_b and D_v decrease with increasing size of the analyzed area (L) because of the increased proportion of the flat part (the part of brittle-type fracture) in the analyzed area, but the values level off above L=300 pixels. The value of D_v is larger than that of D_b also in this case. The dependence of the fractal dimension on the size of the analyzed area can be correlated with microstructures on the fracture surface. Therefore, it is important for the fractal analysis to use an image large enough to involve principal microstructures on the fracture surface.

The fractal dimension of the three-dimensional fracture surface was estimated on relatively wide area involving 8 grains (a Cu-Be alloy, stage I fatigue) to about 500 grains (commercial alumina). Figure 16 shows the relationship between the value of D_b and that of D_v on various fracture surfaces of materials estimated at the same analyzed area of L=360 pixels [34]. The actual size of the analyzed area is equal to or larger than about 4.1×10^{-5} m in this case. The length scale of the fractal analysis is in the range from two pixels to the analyzed area. The stage I fatigue fracture surface has the larger fractal dimension compared with the stage II fatigue fracture surface in a Cu-Be alloy. As described in the previous sections, the stage I fatigue fracture surface is formed for the most part in a ductile manner with slip steps and dimples in a Cu-Be alloy, while the stage II fatigue fracture surface is brittle-type with grain-boundary facets and fine dimples. This suggests that ductile fracture surfaces have the larger fractal dimensions compared with brittle-type fracture surfaces in the same material. However, it is difficult to classify the type of fracture, namely, ductile fracture or brittle-type fracture, in different materials only by the fractal dimension of the fracture surface, since even fractal dimensions of brittle-type fracture surfaces exhibit different values with different materials, depending on the fracture mechanisms. There is a positive correlation between the values of D_b and those of D_v, whereas the values of D_v are always larger than those of D_b, irrespective of fracture mechanisms or fracture patterns. Thus, the value of the fractal dimension depends on the algorithms for the fractal analysis.

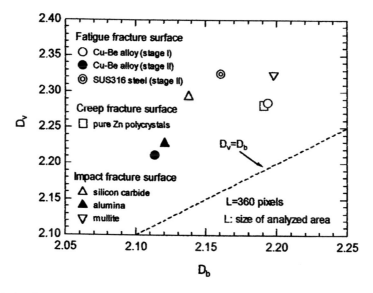

Figure 16 Relationship between the fractal dimension estimated by the box-counting method, D_b, and that estimated by the variance method, D_v, in metals and ceramics (From Tanaka et al., [34]. Reprinted with permission of The Iron and Steel Institute of Japan.).

Two-Dimensional and Three-Dimensional Analyses

The fractal dimension of the fracture surface profile (D_p) was estimated in the length scale range smaller than about one grain-boundary length by the box-counting method. Figure 17 shows examples of the actual (stage I) fatigue fracture surface profiles in the fatigued specimen of a Cu-Be alloy. These micrographs were taken on different specimens [4]. The value of D_p is 1.197 in the parallel direction (Fig. 17a) and 1.180 in the perpendicular direction (Fig. 17b). Similar values of the fractal dimension (i.e. similar extents of the geometrical complexity) of the fracture surface profiles suggest that these fracture surfaces were formed by a common mechanism, namely, a mixed mode of slipping-off in the grains and grain-boundary fracture with dimples [4], whereas the surface roughness (rms) is very different in these two specimens. As shown by arrows, there are overhang (Fig. 17a) and debris (Figs. 17a and 17b) on the fracture surface profile. There are also microcracks linked to the fracture surface (Fig. 17b). These are considered to be microstructural features of the fracture surface [37], which are only partly reproducible by the stereo matching method and may not fully be detected by a scanning tunneling microscope (STM) or an atomic force microscope (AFM).

The surface roughness depended on the profile length and tended to increase with increasing profile length. Further, as shown in Fig. 17, different fracture surface profiles often have very different values of the surface roughness (36.3 μm and 5.67 μm), even if they have the geometrical similarity (the fractal dimensions of the fracture surface profiles are 1.197 and 1.180). Therefore, the morphological features of fracture surfaces cannot be described solely by the surface roughness (rms).

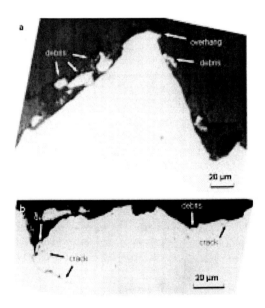

Figure 17 Examples of fracture surface profiles in the fatigued specimen of a Cu-Be alloy; a is a profile in the parallel direction (D_p=1.197, rms=36.3 μm) and b is a profile in the perpendicular direction (D_p=1.180, rms=5.67 μm) (D_p: the fractal dimension of the fracture surface profile; rms: the surface roughness) (From Tanaka et al., [23]. Reprinted with permission of The Iron and Steel Institute of Japan.).

On the contrary, the fractal dimension of the fracture surface profiles can characterize the geometrical features of the fatigue fracture surface in the length scale range smaller than one grain-boundary length (about 14 μm), which is associated with the size range of characteristic microstructures such as slip steps and dimple patterns on the fracture surface [2, 4]. Thus, the fractal dimension of the contours and that of the fracture surface profiles are considered to be a useful index not only for characterizing fracture surfaces but also for comparison of the result of three-dimensional image reconstruction with the actual fracture surface morphology.

The contours of the fracture surface were extracted from the reconstructed three-dimensional images in a Cu-Be alloy [23]. Figure 18 shows the contours used for fractal analysis imposed on the height image of the stage I fatigue fracture surface in a Cu-Be alloy. The numbers in the figure indicate the relative height. The fractal dimension of the contours superimposed on the height image was estimated on representative 13 contours [23]. Figure 19 shows the fractal dimension of thirteen contours (Fig. 17) in the reconstructed three-dimensional image. The fractal dimension of the contours was estimated in the scale length range smaller than about one grain-boundary length (about 14 μm) [23]. The fractal dimension lies between about 1.17 and about 1.29, and the mean value of the fractal dimension is about 1.238. Also shown in the figure is the fractal dimension of the actual fracture surface profiles in two directions. The mean value of the fractal dimension of the actual fracture surface profiles was about 1.210 in the parallel direction [4] and about 1.190 in the perpendicular direction. The fractal dimension of the contours obtained in this study is close to that of the fracture surface profiles. Thus, the stereo matching method can reproduce the three-dimensional image of the complex fracture surface. The fatigue fracture surface in the Cu-Be alloy seems to be isotropic and self-similar at least in the length scale range smaller than about one grain-boundary length.

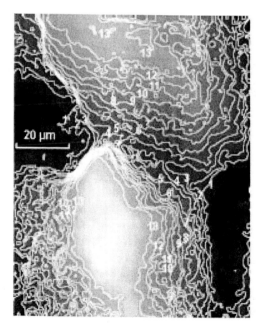

Figure 18 The contours used for fractal analysis in the height image of the stage I fatigue fracture surface imposed on a Cu-Be alloy (numbers indicate the relative height) (From Tanaka et al. , [23]. Reprinted with permission of The Iron and Steel Institute of Japan.).

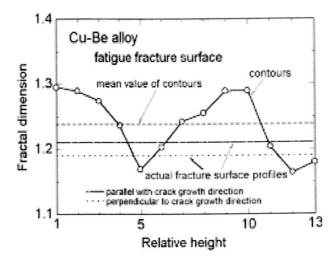

Figure 19 The fractal direction of thirteen contours (Fig. 17) extracted from the reconstructed three-dimensional image of the fatigue fracture surface in a Cu-Be alloy (From Tanaka et al., [23]. Reprinted with permission of The Iron and Steel Institute of Japan.).

However, the fracture surface profiles extracted from the reconstructed image were found to be a little simpler in detail than the actual fracture surface profiles in the fatigued specimens of a Cu-Be alloy, while the reconstructed image reproduces principal features of the actual fracture surface [23]. The mean value of the fractal dimension of the reconstructed fracture surface profile was about 1.128 and was smaller than that of the actual fracture surface profiles (1.210) in the parallel direction [23]. The mean value of the fractal dimension was about 1.146 and was a little smaller than that of the actual fracture surface profiles (1.190) also in the perpendicular direction. As described above, actual fracture surface involves overhang and debris in addition to microcracks linked to the fracture surface (Fig. 17). These microstructural features are only partly reproducible by the stereo matching method. This is why a smaller value of the fractal dimension was obtained on the fracture surface profiles extracted from the reconstructed three-dimensional image.

Table 1 lists the fractal dimension of the three-dimensional fracture surface, D_b, and that of the fracture surface profile or the indentation crack profile, D_p, in metals and ceramics [34]. The size of the analyzed area in the three-dimensional fractal analysis is equal to or larger than about 6.5×10^{-5} m. Both fractal dimensions were estimated in the length scale range smaller than the grain size or smaller than about one grain-boundary length, since the sizes of microstructures such as slip steps, dimples and grain-boundary steps on the fracture surfaces are smaller than the grain size or one grain-boundary length. About 8 grains were involved in the analyzed area (360x360 in pixel) of the stage I fatigue fracture surface in the Cu-Be alloy, while about 10 grains were included in the analyzed area (600x600 in pixel) of the stage II fatigue fracture surface. The analyzed area of the stage II fatigue fracture surface (360x360 in pixel) involved about 25 grains in the SUS316 steel. Much larger number of grains seemed to be involved in the fracture surfaces of the ceramics. Thus, the three-dimensional fractal analysis was carried out in relatively wide areas of fracture surfaces in this study. In a Cu-Be alloy, the length scale range of the fractal analysis is associated with the size range of slip steps in the stage I fatigue fracture surface (about 6×10^{-7} m to about 5×10^{-6} m), and is also related to the size range of dimples (about 3.5×10^{-7} m to about 9.0×10^{-6} m) or that of grain-boundary steps (about 4×10^{-6} m to about 3×10^{-5} m) in both the stage I and stage II fatigue fracture surfaces [4]. The length scale range of the fractal analysis is associated with the striation spacing (about 5×10^{-7} m to about 6×10^{-6} m) in the stage II fatigue fracture surface of the SUS316 steel [4]. Therefore, the fractal dimension can be correlated with specific microstructures in materials.

Table 1 The fractal dimension of the three-dimensional fracture surface, D_b, and that of the fracture surface profile or the indentation crack profile, D_p, in metals and ceramics (From Tanaka et al., [34]. Reprinted with permission of The Iron and Steel Institute of Japan.).

Materials	Conditions of fracture surface	Average grain size (GS) or grain-boundary length (GBL) (grain size range) (m)	D_b (length scale range for fractal analysis, m)	D_p (length scale range for fractal analysis, m)	
Cu-Be alloy	Stage I fatigue	GS=2.4×10^{-5} GBL=1.4×10^{-5}	2.164 (3.6×10^{-7} to 1.3×10^{-5})	P	1.210 (6.7×10^{-7} to 1.7×10^{-5})
				T	1.190 (6.7×10^{-7} to 1.7×10^{-5})
	Stage II fatigue		2.116 (2.4×10^{-7} to 1.4×10^{-5})	P	1.168 (6.7×10^{-7} to 1.7×10^{-5})
SUS316 steel	Stage II fatigue	GS=1.3×10^{-5} GBL=8×10^{-6}	2.198 (3.6×10^{-7} to 8.1×10^{-6})	P	1.221 (6.7×10^{-7} to 8.3×10^{-6})
Silicon carbide (Norton NC-430)	Impact fracture	GS=1.3×10^{-5} (3.7×10^{-7} to 8.6×10^{-5})	2.189 (2.3×10^{-7} to 5.7×10^{-6})		1.16* (2.7×10^{-7} to 5.5×10^{-6})
Alumina (SSA-H)	Impact fracture	GS=3.0×10^{-6} (6.4×10^{-7} to 9.2×10^{-6})	2.134 (3.0×10^{-7} to 8.9×10^{-6})		1.15** to 1.19**

P: in the plane in parallel with the crack growth direction [4]; T: in the plane transverse to the crack growth direction; *: the value for indentation crack [32];,**: the values for indentation cracks in similar alumina [32].

The fractal dimension of a surface can be predicted by adding unity to the fractal dimension of its cross-section from general properties of fractal sets [38]. Figure 20 shows the relationship between the fractal dimension of the three-dimensional fracture surface, D_b, and the fractal dimension, D_p+1, predicted by the two-dimensional fractal analysis in metals and ceramics [34]. The length scale range of the fractal analysis is shown in Table 1. As described in section 5.1, the fracture mechanisms and fracture patterns are quite different in metals and ceramics. However, the stage I (ductile) fatigue fracture surface has the larger value of the fractal dimension compared with the stage II (brittle-type) fatigue fracture surface in a Cu-Be alloy. The value of D_b is close to that of D_p+1 in materials examined in this study, although the former value is slightly smaller than the latter value in most cases. This may indicate that principal fracture surface patterns like slip steps, dimples or micropores can be reproduced by the stereo matching method. However, there seems to be other microstructural features like overhang and debris generated by crack overlapping and microcracks linked to the fracture surfaces [37], which are only partly reproducible by the stereo matching method and may not fully be detected by scanning probe microscopes. This may be why the value of D_b is slightly smaller than that of D_p+1 in these materials.

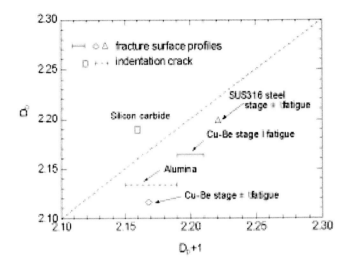

Figure 20 Relationship between the fractal dimension of the three-dimensional fracture surface, D_b, and the fractal dimension, D_p+1, predicted by the two-dimensional fractal analysis in metals and ceramics (From Tanaka et al., [34]. Reprinted with permission of The Iron and Steel Institute of Japan.).

Kobayashi and Shockey [19] revealed that the fracture toughness for crack initiation (J_{IC}) estimated from the reconstructed fracture surface profiles coincided with the value obtained using a conventional fracture test procedure in steels. Further, Stampfl and Kolednik [21] correlated the local fracture toughness estimated from the reconstructed images with the specific microstructures. Their results also show that the three-dimensional image reconstruction using the stereo matching method can reproduce the principal features of fracture surfaces and is a useful tool for the estimation of fracture toughness values or for the study of micro fracture mechanisms in materials. Two-dimensional analysis of actual fracture surface profiles can reveal microstructures such as debris, overhang and microcracks linked to the fracture surface, which cannot be detected by three-dimensional fracture surface analysis [2-4, 8, 39]. However, the geometrical information about a given fracture surface in a wide area can be obtained by the three-dimensional fracture surface analysis [18-23, 34, 40]. Combination of three-dimensional and two-dimensional analyses may lead to the further understanding of the geometrical features of fracture surfaces and the fracture mechanism in materials.

Three-Dimensional Fractography by Shape Parameter Mapping

As described above, the fractal dimension and the surface roughness are very different shape parameters. The fractal dimension describes the complexity and self-similarity of object shape, and has a scale-independence, while the surface roughness measures the averaged magnitude of undulation on the surface. A combination of fractal dimension and surface roughness may give important information about fracture surface patterns which cannot be detected by conventional method, since these shape parameters can also represent the local variation of the fracture surface patterns. The fractal dimension map (FDM) and the surface roughness map (SRM) (displayed in 256 grey scale levels) composed from the height data are proposed for the investigation of the local fracture mechanisms and the fracture process in materials.

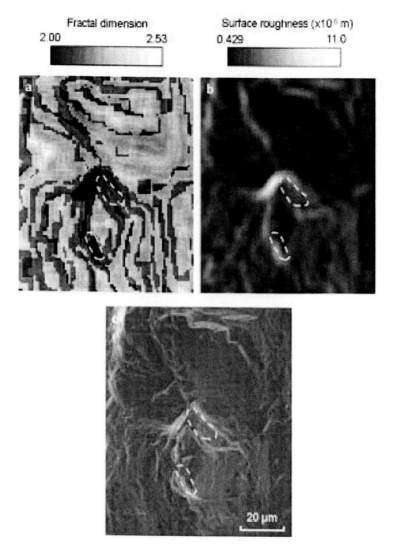

Figure 21 Stage I fatigue fracture surface of a Cu-Be alloy; a is FDM; b is SRM (the calculated area (m x m) is 24 x 24 in pixel and the displayed area (k x k) is 4 x 4 in pixel in a and b) and c is original SEM image (403 x 467 in pixel) (The regions where grain-boundary fracture occurred are enclosed by broken lines.)

Figure 21 shows the original SEM image, FDM and SRM on the stage I fatigue fracture surface of a Cu-Be alloy. The calculated area is 24 pixels x 24 pixels and the displayed area is 4 pixels x 4 pixels in FDM (Fig. 21a) and SRM (Fig. 21b). The value ranges of the fractal dimension and the surface roughness are also shown in the figure. The brighter part in grey scale level shows the region with the relatively larger value of the fractal dimension (in FDM) or the surface roughness (in SRM). The stage I fatigue fracture surface was formed for the most part in a ductile manner (Fig. 21c) [23]. The part of ductile fracture is indicated by the bright region in FDM, while this part does not always show a bright contrast in SRM. The flat part with the fractal dimension of around 2.17 (dark regions in both FDM and SRM and indicated by broken lines in the figure) may be formed by grain-boundary fracture, since the fractal dimension of the fatigue fracture surface profile is 1.168 on the stage □ fatigue fracture surface of the same alloy where grain-boundary fractures prevails [4].

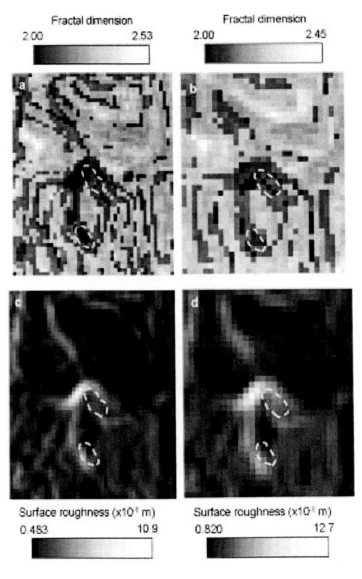

Figure 22 FDM and SRM on the stage I fatigue fracture surface of a Cu-Be alloy (grain-boundary facets are enclosed by broken lines); a is FDM (m=24, k=8 in pixel); b is FDM (m=36, k=12 in pixel); c is SRM (m=24, k=8 in pixel) and d is SRM (m=36, k=12 in pixel)

The flat part is also shown by broken lines in the maps composed under different conditions of sizes of calculated area (m) and displayed area (k) (Fig. 22). Principal features of FDM and SRM do not significantly change with conditions of calculation and display. The decrease of the displayed area enabled the detection of local fracture patterns using FDM and SRM (compare Figs. 21a and 21b with Fig. 22), although some patterns become unclear with decreasing displayed area (k). The increase of the calculated area (m) reduced noises in the maps, especially in FDM. However, both increases of the size of calculated area (m) and that of displayed area (k) result in unclear fracture patterns (Figs. 22b and 22d). The increase of the calculated area (m) led to the increase of the value range of the surface roughness and to the slight decrease of the value range of the fractal dimension. Steeply inclined part also shows a dark contrast in FDM, but this part is very bright in SRM, irrespective of local fracture mechanisms.

Figure 23 shows the original SEM image, FDM and SRM on the impact fracture surface of a silicon carbide. A local variation of fracture patterns can be observed in both maps (Figs. 23c and 23d). Areas of bright contrast in FDM (large fractal dimension) and intermediate contrast in SRM correspond to regions of complex geometry. There are "river-like" patterns, which are dark in FDM (Fig. 23c) and show a bright contrast in SRM (Fig. 23d), and these are considered as steeply inclined parts. As shown in the schematic illustration (Fig. 23b), these "rivers" seem to join at some places and the local crack growth direction can be known from these patterns (indicated by an arrow in Fig. 23b). Further, there are relatively flat areas, which show a dark contrast in both FDM (Fig. 23c) and SRM (Fig. 23d). Some of them are illustrated in Fig. 23b, and are considered as the regions where brittle-type fracture occurred [34]. These characteristic patterns cannot be detected in the original SEM image (Fig. 23a). As marked by broken lines in Fig. 24, these features seem to be retained even when mapping conditions, namely, the size of calculated area (m) and displayed area (k) are changed in both FDM and SRM, although changes in mapping conditions may lead to the extinction and emergence of some local patterns. Therefore, it is necessary to choose proper mapping conditions of FDM and SRM for extraction of characteristic patterns on the fracture surfaces of materials. Similar "river-like" patterns and their joining are observed on the impact fracture surface of an alumina (Fig. 25). By comparing FDM with SRM, one can discriminate the areas of complex geometry, the flat areas and the steeply inclined parts. "River-like" patterns are much clearer in FDM (Fig. 25c) than in SRM (Fig. 25d) in this case. Joining of "rivers" is visible at some places and the local crack growth direction (approximately from top to bottom) can be known as schematically illustrated in Fig. 25b, although the fracture origin does not seem to exist in the original SEM image (Fig. 25a). Characteristic patterns such as "river-like" patterns are not visible in the original micrograph (Fig. 25a) [34]. It is worth noting that FDM and SRM can detect the fracture surface patterns that cannot be observed only by microscopes. Thus, one may know the local fracture mechanisms, the crack growth direction or the fracture origin in a given fracture surface using FDM and SRM. This new method of shape parameter mapping can detect characteristic patterns on a fracture surface, which cannot be observed solely with a scanning electron microscope, and may lead to a new research field, namely, three-dimensional fractography.

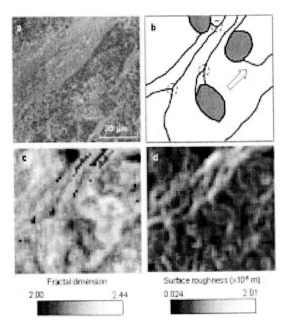

Figure 23 Impact fracture surface of a silicon carbide (Norton NC-430); a is original SEM image (663 x 601 in pixel); b is schematic illustration (lines show river-like patterns, marked areas are the regions where "brittle fracture" occurred, and an arrow indicates the local crack growth direction (joining of "rivers" is shown by broken circles)); c is FDM; d is SRM (m=36, k=12 in pixel in c and d)

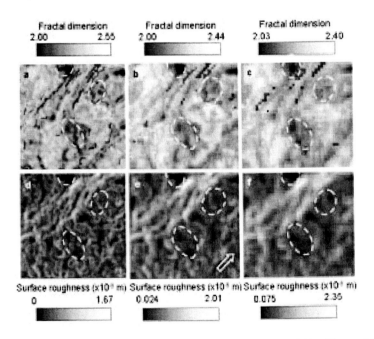

Figure 24 FDM and SRM on the impact fracture surface of a silicon carbide (Norton NC-430); a is FDM (m=24, k=8 in pixel); b is FDM (m=36, k=12 in pixel); c is FDM (m=48, k=16 in pixel); d is SRM (m=24, k=8 in pixel); e is SRM (m=36, k=12 in pixel) and f is SRM (m=48, k=16 in pixel) (an arrow shows the local crack growth direction and broken circles show the regions of brittle fracture.)

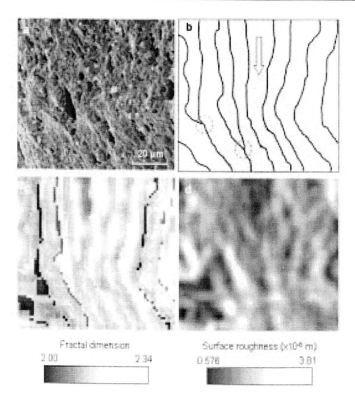

Figure 25 Impact-fractured surface of an alumina (SSA-H); a is original SEM image (575 x 513 in pixel); b is schematic illustration (lines indicate "river-like" patterns and an arrow shows the main crack growth direction (joining of "rivers" is shown by broken circles); c is FDM; d is SRM (the calculated area is 48 x 48 in pixel and the displayed area is 8 x 8 in pixel in c and d)

Features and Application of Shape Parameter Mapping

As described in the previous section (5.4), there seems to be several kinds of common microstructural features on the fracture surfaces of materials. The characteristic microstructures can be detected by mapping shape parameters, namely, the fractal dimension and the surface roughness, on the fracture surfaces. The relative values of the fractal dimension or the surface roughness are displayed in 256 grey scale levels in the fractal dimension map (FDM) or in the surface roughness map (SRM). The brighter part in the map indicates the region of the larger value of the fractal dimension or the surface roughness.

A schematic illustration shows how a fracture surface can be observed by tilting (Fig. 26). By considering the direction of observation on the fracture surface, it is clear that ledges and bumps on the original fracture surface (Fig. 26a) are only partly detectable on the steeply inclined part (Fig. 26b), and the apparent fracture surface is very simple (Fig. 26c). The height difference on the fracture surface becomes larger by tilting whatever the original shape of the fracture surface is (Fig. 26c), and this leads to a larger value of the surface roughness. Thus, a steeply inclined part generally shows a smaller value of the fractal dimension and a larger value of the surface roughness, irrespective of the original surface geometry. This part is shown as a dark region in FDM and as a bright region in SRM. However, these patterns in

FDM or SRM give a clue for investigation of the local fracture mechanisms, the crack growth direction or the fracture origin in a given fracture surface. Pattern extraction by mapping shape parameters (FDM and SRM) enables "hidden patterns" on fracture surfaces to be interpreted in terms of fractography.

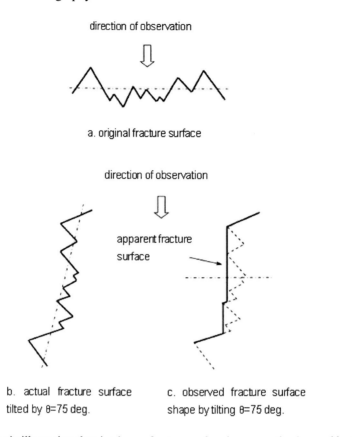

Figure 26 Schematic illustration showing how a fracture surface is apparently observed by tilting.

According to the analytical results on the fracture surfaces of materials, the characteristic fracture surface patterns are displayed as follows in the fractal dimension map (FDM) and the surface roughness map (SRM) (Table 2) [41]:

1. An area that shows bright contrast in FDM but not always bright contrast in SRM, corresponds to a region of relatively complex geometry such as a ductile fracture surface.
2. A region that shows dark contrast in both FDM and SRM, corresponds to a relatively flat region such as a brittle fracture surface.
3. An area that shows dark contrast in FDM and bright contrast in SRM, corresponds to a steeply inclined part like a step. This part apparently shows the smaller fractal dimension and the larger surface roughness, irrespective of fracture mechanisms.

Table 2 Characteristic fracture surface patterns displayed in the fractal dimension map (FDM) and the surface roughness map (SRM) [41].

fracture surface patterns	FDM	SRM
region of relatively complex geometry such as a ductile fracture surface	bright	not always bright
relatively flat region such as a brittle fracture surface	dark	dark
steeply inclined part like a step	dark	bright

It should be noted that these three features in FDM and SRM are non-overlapping in any cases. Therefore, local fracture patterns can be visually known from inspection of these features on FDM and SRM.

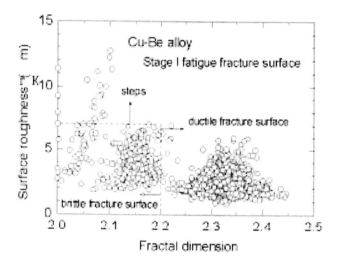

Figure 27 The relationship between the fractal dimension and the surface roughness in small regions of the stage I fatigue fracture surface in a Cu-Be alloy. (the analysed fracture surface is 372 x 432 in pixel, and the condition of the calculation is m=36 and k=12 in pixel) [41].

Figure 27 shows the relationship between the fractal dimension and the surface roughness in small regions of the stage I fatigue fracture surface in a Cu-Be alloy (Fig. 4) [41]. The analyzed area of the fracture surface (372 x 432 in pixel and almost the same as Fig. 4c) was divided into 1116 small regions. Relatively large values of the size of the calculated area (m=36 pixels) and that of the small regions (k=12 pixels) were chosen in the calculation (Figs. 22b and 22d) in order to extract principal features of local fracture surface patterns in FDM and SRM in this case. In the Cu-Be alloy, the fractal dimension of the fracture surface profile (Dp, 1<Dp<2) was in the range from 1.190 to 1.210 for the stage I fatigue fracture surface in which the ductile fracture mechanism prevailed [23], whereas the value of Dp was 1.168 for the stage II fatigue fracture surface formed by grain-boundary fracture (brittle fracture) [4]. It is assumed in this study that an area in which the fractal dimension of the three-dimensional fracture surface (Db, 2<Db<3) estimated by the box-counting method is less than 2.20, belongs to the brittle fracture region, while an area in which the value of the surface roughness (rms) is more than about 7.0 μm, is regarded as a step. The brittle fracture surface was formed by grain-boundary fracture in the Cu-Be alloy. The proportion of the area is 70.1 % for the ductile fracture surface, 27.9 % for the brittle fracture and 2.0 % for steps. If the steps are excluded in the estimation, the area proportion of the brittle fracture surface slightly increases to 28.5 % (that of the ductile fracture surface is 71.5 %).

If all the steeply inclined parts on the fracture surface are assumed to be formed in a ductile manner, for example, by shear stress (anti-plane shear mode, mode III) [42], these parts may be classified into the ductile fracture surface. The area proportion of the grain-boundary fracture surface gives the minimum value (27.9 %) in this case. The steeply inclined parts on the fracture surface may be formed on grain boundaries in a brittle manner. If these parts are assumed to belong to the grain-boundary facets, the area proportion of the brittle fracture surface gives the maximum value, 29.9%, for the stage I fracture surface in this case. Some of the steeply inclined parts may belong to the ductile fracture surface, while others should be classified into the brittle fracture surface. The true area proportion of the brittle fracture surface may lie between the minimum value (27.9 %) and the maximum value (29.9 %). These results qualitatively coincide with the microscopic observation that the stage I fatigue fracture surface was formed for the most part in a ductile manner with slip steps and dimples except small grain-boundary facets in the Cu-Be alloy. In general, it is necessary to choose the moderate calculation condition and the exact threshold values of the fractal dimension and the surface roughness for classification of different fracture patterns (for example, ductile fracture surface, brittle fracture surface and step). Thus, the geometrical analysis by shape parameters can be applied not only to the detection of characteristic fracture patterns but also to the quantitative evaluation of the specific fracture surfaces in materials.

Conclusions

Three-dimensional fracture surfaces of metals and ceramics produced by different mechanisms were reconstructed by a new stereo matching method based on the coarse-to-fine format. The fractal dimension of the three-dimensional fracture surface was estimated by the box-counting method using the height data generated by the stereo matching method. Discussion was made on the dependence of the fractal dimension of the fracture surface on the microstructures of materials, the methods of the fractal analysis and the size of the analyzed area. The results of the three-dimensional fractal analysis were compared with those obtained by the two-dimensional fractal analysis. Further, a new method of mapping shape parameters, namely, the fractal dimension and the surface roughness, was developed for quantitative three-dimensional fractography of fracture surfaces in materials. The results obtained were summarized as follows.

(1) The mean value of the fractal dimension of the contours in the stage I fatigue fracture surface of a Cu-Be alloy was about 1.238, and was close to the values of the actual fracture surface profiles in the plane in parallel with the crack growth direction (about 1.210) and in the plane perpendicular to the crack growth direction (about 1.190). Thus, the stereo matching method can reproduce the principal three-dimensional features of the complex fracture surface. The fatigue fracture surface in the Cu-Be alloy seems to be isotropic at least in the length scale range smaller than about one grain-boundary length. However, there were

microstructural features such as overhang, debris and microcracks on the fatigue fracture surface of a Cu-Be alloy, which were only partly reproducible by the stereo matching method.

(2) There was a good correlation between the results of the three-dimensional fractal analysis and those of the two-dimensional fractal analysis on the fracture surfaces of materials. The three-dimensional fractal analysis by the box-counting method is essentially applicable to any surfaces, if the information in the height direction is obtained. The geometrical information about a given fracture surface in a wide area can be obtained by the three-dimensional fracture surface reconstruction and analysis. A combination of three-dimensional and two-dimensional analyses may lead to the further understanding of the geometrical features of fracture surfaces and the fracture mechanism in materials.

(3) The value of the fractal dimension of the three-dimensional fracture surface depends not only on the size of the analyzed area but also on the algorithms of the fractal analysis. The fractal dimension estimated by the variance method was larger than that evaluated by the box-counting method, irrespective of fracture mechanisms of materials. The fractal dimension of the three-dimensional fracture surface as well as that of the fracture surface profile, which are estimated in a given length scale range can be correlated with specific microstructures like slip steps or dimple patterns on the fracture surface. The fractal dimension is a useful index not only for characterizing fracture surfaces but also for comparison of the result of three-dimensional image reconstruction with the actual fracture surface morphology.

(4) In a Cu-Be alloy, the ductile fracture surface (stage I fatigue) had the larger fractal dimension compared with the brittle-type fracture surface (stage II fatigue) in a Cu-Be alloy. However, it was difficult to classify the type of fracture, namely, ductile fracture or brittle-type fracture, in different metallic materials and ceramics only by the fractal dimension of the fracture surface, since even fractal dimensions of brittle-type fracture surfaces exhibited different values with different materials, depending on the fracture mechanisms of materials.

(5) The fractal dimension is a scale-independent shape parameter, while the surface roughness depends on the actual length scale. By mapping these two different shape parameters as the fractal dimension map (FDM) and the surface roughness map (SRM), one can detect the "hidden" fracture patterns that cannot be recognized by conventional fractography. FDM and SRM could detect the regions of ductile fracture, those of grain-boundary fracture and the steeply inclined parts on the fracture surface. For example, "river-like" patterns or the regions of brittle-type fracture, which was not observed in the original scanning electron microscope (SEM) image, were recognized using FDM and SRM on the impact fracture surfaces of a SiC and an alumina. Pattern recognition using mapping technologies of FDM and SRM is applicable not only to the investigation of the local fracture mechanisms, the crack growth direction or the fracture origin but also to the quantitative evaluation of the area proportions of the specific fracture surfaces in materials.

Acknowledgements

The authors thank The Iron and Steel Institute of Japan (ISIJ) and Mitsutoyo Association for Science and Technology (MAST) for financial support.

References

[1]   B.B. Mandelbrot, D.E. Passoja and A.J. Paullay, Nature, **308** (1984), 721.

[2]   R.H. Dauskardt, F. Haubensak and R.O. Ritchie, Acta Metall. **38** (1990), 142.

[3]   M. Tanaka, J. Mater. Sci., **27** (1992), 4717.

[4]   M. Tanaka, A. Kayama and R. Kato, J. Mater. Sci. Lett., **18** (1999), 107.

[5]   T. Ikeshoji and T. Shioya, Fractals, **7** (1999), 159.

[6]   X.W. Li, J.F. Tian, S.X. Li and Z.G. Wang, Materials Transactions, **42** (2001), 128.

[7]   A.M. Gokhale, W.J. Drury and F. Mishra, ASTM STP1203, edited by J.E. Masters and L.N. Gilbertson (American Society for Testing and Materials, Philadelphia, PA, 1993) p. 3.

[8]   Z.G. Wang, D.L. Chen, X.X. Jiang, S.H. Ai and C.H. Shih, Scripta Metall., **22** (1988), 827.

[9]   S. Matsuoka, H. Sumiyoshi and K. Ishikawa, Trans. Japan Soc. Mech. Eng., **56** (1990), 2091.

[10] E.E. Underwood and K. Banerji, Mater. Sci. Eng., 80 (1986), 1.

[11] M. Tanaka, Z. Metallkd., **88** (1997), 217.

[12] N. Almqvist, Surface Science, **355** (1996), 221.

[13] V.Y. Milman, N.A. Stelmashenko and R. Blumenfeld, Progress in Mater. Sci., **38** (1994), 425.

[14] P. Streitenberger and D. Förster, Physica Status Solidi, **B 171** (1992), 21.

[15] M. Takahashi and H. Hasegawa, Fractals, **8** (2000), 189.

[16] M. Tanaka and H. Iizuka, Z. Metallkd., **82** (1991), 442.

[17] P. Streitenberger, D. Förster, G. Kolbe and P. Veit, Scripta Metall. Mater., **33** (1995), 541.

[18] K. Komai and J. Kikuchi, J. Soc. Mater. Sci. Japan, **34** (1985), 648.

[19] T. Kobayashi and D.A. Shockey, Metall. Trans., **18A** (1987), 1941.

[20] J. Stampfl, S. Scherer, M. Berchthaler, M. Gruber and O. Kolednik, Int. J. Fracture, **78** (1996), 35.

[21] J. Stampfl and O. Kolednik, Int. J. Fracture, **101** (2000), 321.

[22] Y. Kimura, M. Tanaka and R. Kato, ISIJ International, **44** (2004), : Y. Kimura and M. Tanaka, Proceedings of the Fourth International Conference on Materials for Resources, Akita, Japan, October 11-13, 2001, Vol. 2, pp. 249-253.

[23] M. Tanaka, Y. Kimura, L. Chouanine, J. Taguchi and R. Kato, ISIJ International, **43** (2003), 1453.

[24] H. Takayasu, "Fractals in the Physical Sciences", (Manchester University Press, Manchester and New York, 1990), p. 6.

[25] M. Tanaka, J. Mater. Sci., **31** (1996), 749.

[26] B.B. Mandelbrot, "The Fractal Geometry of Nature", translated by H. Hironaka, (Nikkei Science, Tokyo, 1985), p. 25, p. 108.

[27] D. Sen, S. Mazumder and S. Tarafdar, J. Mater. Sci., **37** (2002), 941.

[28] U. Wendt, K. Stiebe-Lange and M. Smid, J. Microscopy, **207** (2002), 169.

[29] M. Tanaka and A. Kayama, J. Mater. Sci. Lett., **20** (2001), 907.

[30] Kayama, M. Tanaka and R. Kato, J. Mater. Sci. Lett., **19** (2000), 565.

[31] M. Tanaka, Y. Kimura, L. Chouanine, R. Kato and J. Taguchi, J. Mater. Sci. Lett., **22** (2003), 1279.

[32] M. Tanaka, J. Soc. Mater. Sci. Japan, **47** (1998), 169.

[33] P.-S. Chen and R.C. Wilcox, Mater. Characterization, **26** (1991), 9.

[34] M. Tanaka, Y. Kimura, A. Kayama, R. Kato and J. Taguchi, ISIJ International, **44** (2004), 1250.

[35] M. Tanaka, A. Kayama, Y. Sato and Y. Ito: J. Mater. Sci. Lett., **17** (1998), 1715.

[36] M. Tanaka, R. Kato and A. Kayama, J. Mater. Sci., **37** (2002), 3945.

[37] D. Hull, "Fractography", (Cambridge University Press, Cambridge, 1999), p. 118.

[38] H. Takayasu, "Fractals in the Physical Sciences", (Manchester University Press, Manchester and New York, 1990), p. 20.

[39] M. Tanaka, Z. Metallkd., **84** (1993), 697.

[40] J. Jun and S. Sakai, J. Soc. Mater. Sci. Japan, **50** (2001), 1176.

[41] M. Tanaka, Y. Kimura, R. Kato and N. Oyama, submitted to J. Mater. Sci. Lett.

[42] D. Hull, "Fractography", (Cambridge University Press, Cambridge, 1999), p. 91.

In: Trends in Materials Science Research
Editor: B.M. Caruta, pp. 81-99

ISBN: 1-59454-367-4
© 2006 Nova Science Publishers, Inc.

Chapter 3

OPTICAL PROPERTIES OF PD-FREE AU-PT-BASED HIGH NOBLE ALLOYS

Takanobu Shiraishi[1]

Division of Dental and Biomedical Materials Science, Department of Developmental
and Reconstructive Medicine, Graduate School of Biomedical Sciences,
Nagasaki University, 1-7-1 Sakamoto, Nagasaki 852-8588, Japan

Abstract

The effects of the addition of Pt to Au on its optical properties and the effects of small
additions of various base metals to a parent binary Au-10 at.% Pt alloy on its optical
properties were investigated by means of spectrophotometric colorimetry. Spectral reflectance
curves for the mirror-polished flat samples were collected in the wavelengths ranging from
360 to 740 nm. Three-dimensional color coordinates, *i.e.* L^*, a^*, and b^*, in the CIE
(Commission Internationale de l'Eclairage) 1976 $L^*a^*b^*$ (CIELAB) color space were
obtained to specify the color of the sample.

The strong decolorizing effect of Pt was evidenced when Au was alloyed with Pt. This
decolorizing effect was considered to be due to the formation of "virtual bound states". The
alloying addition of a small amount of base metals with a high number of valence electrons to
the Au-10 at.% Pt alloy effectively gave a gold tinge to the parent Au-10 at.% Pt alloy.
Analysis of spectral reflectance curves for all the alloys revealed that the position of the
absorption edge in the visible spectrum was not affected by the alloying elements within the
limit of experimental error. However, the slope of the spectral reflectance curve at its
absorption edge near 515 nm (approximately 2.4 eV) systematically increased with increasing
number of valence electrons per atom, e/a, in the alloy. As a result, with increasing e/a-value a
chromaticity index b^* (yellow-blue coordinate) markedly increased and a chromaticity index
a^* (red-green coordinate) slightly increased, giving a gold tinge to the parent Au-Pt alloy. It
was found that the number of valence electrons per atom in an alloy plays an important role in
determining color of Au-Pt-based high noble alloys. This information is expected to be useful
in controlling the color of Pd-free Au-Pt-based high noble alloys.

---

[1] E-mail address: siraisi@net.nagasaki-u.ac.jp, Telephone: +81-95-849-7659, Fax: +81-95-849-7658

Introduction

Porcelain-fused-to-metal (hereafter referred to as PFM) restorations have become a widespread technology in dentistry [1, 2]. There are several compositional types of precious-metal PFM alloys, such as Au-Pt-Pd, Au-Ag-Pd, Au-Pd, and Pd-Ag systems [3]. Among them, the Au-Pt-Pd-based high noble alloys have the advantage that they have been around for some considerable time and clinical experience has shown that they are extremely successful [4]. In particular, the bond between the ceramic and the metal is very strong and highly reliable [4].

Recently, Pd-free Au-Pt-based high noble dental alloys were developed and introduced into the market, probably due to a recent sharp rise in the price of palladium [5]. These Pd-free Au-Pt-based high noble alloys can be used not only as fixed crown- and bridgework for porcelain veneers but also as materials for CAD/CAM, full cast crown, and bridgeworks. This new type of high noble alloy may be attractive to patients who are allergic or hypersensitive to palladium. Although the primary concerns about cast dental restorations are with biocompatibility, and corrosion and tarnish resistance, it would be naïve to discount the strong emotional attachment to selected alloy colors involving gold [6]. In fact, German [7] pointed out that the Au-Pd-based alloys for PFM restorations tend to be very light in color and have not been universally accepted in spite of the very attractive properties. Instead, the attraction of traditional high-gold PFM alloys is with the light-yellow color [7]. Thus, the author of the present article considers that color control is one of the important criteria in developing dental precious-metal alloys containing gold.

Historically, the color of an alloy was expressed subjectively while mechanical and physical properties have been treated quantitatively [6]. For instance, Leuser [8] investigated the relationship between color and chemical composition in the ternary Au-Ag-Cu system. Based on an extensive visual examination of the prepared alloys, he divided the whole range of the Au-Ag-Cu system into several areas according to color. Later, the color of an alloy was expressed with three-dimensional color coordinates, *i.e.* L^*, a^*, and b^*, in the CIELAB color space. These color coordinates were obtained by using a colorimeter. However, in order to understand the mechanism of color change with chemical composition, observation and analysis of spectral reflectance curves in the visible spectrum is important. The recent development of a computer-controlled spectrophotometer has made this possible.

The recent Pd-free Au-Pt-based high noble alloys for PFM restorations are composed of Au and Pt with small amounts of base metals such as In, Fe, Zn, Sn, and others. As in the cases of the Au-Pt-Pd-based high noble alloys, these base metals are added to the Pd-free Au-Pt-based high noble alloys as hardening ingredients [1, 9, 10] and for their ability to bond porcelain to the metal surface [2]. Nobility (total content of Au and Pt group metals) of these Au-Pt-based high noble alloys is extremely high at as much as 97 to 98%. It is interesting to note that although the Au and Pt contents in this type of commercial alloy are almost always the same, *i.e.* 86-87% Au and 10-11% Pt, the color of an alloy slightly varies depending on the elements and contents of the above-mentioned base metals. This fact suggests that the color of a Au-Pt-based high noble alloy is affected by small quantities of base metals such as In, Fe, Zn, Sn, and others.

This article reports the effects of the alloying additions of various base metals with a different number of valence electrons on the optical properties of the parent Au-Pt alloy.

Mechanism of color change with chemical composition in the Pd-free Au-Pt-based high noble alloys was investigated.

CIELAB Color Space

Numerical Expression of the Color of an Object

The color of an object can be expressed in terms of three-dimensional coordinates in the uniform color scale. In this three-dimensional uniform color space, the color of an object is defined by lightness, L^*, red-green chromaticity index, a^*, and yellow-blue chromaticity index, b^*, as shown in Fig. 1. The L^*-value can range from 0 (black) to 100 (white). A positive value of a^* coordinate corresponds to red and a negative value of a^* coordinate corresponds to green. A positive value of b^* coordinate corresponds to yellow and a negative value of b^* coordinate corresponds to blue. It is noted that the intersection of a^* and b^* axes corresponds to gray. The three coordinates in the CIELAB color space, L^*-, a^*-, and b^*-values, can be obtained as direct readouts from a recent computer-controlled spectrophotometer.

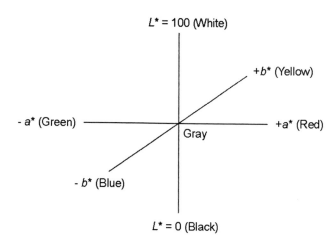

Fig. 1 Three-dimensional color coordinates in the CIE 1976 $L^*a^*b^*$ (CIELAB) uniform color space.

Color Difference between Two Objects

One of the advantages of using the CIELAB uniform color space in expressing the color of the object is that the color difference between two objects can be expressed by a simple parameter, ΔE^*. The ΔE^*-value is the distance between two points, each expressed in terms of L^*, a^*, b^*, in the CIELAB uniform color space, as shown in Fig. 2. The color difference parameter ΔE^*-value is then calculated by the following expression:

$$\Delta E^* = [(\Delta L^*)^2 + (\Delta a^*)^2 + (\Delta b^*)^2]^{1/2} \tag{1}$$

where,

$$\Delta L^* = L^*_1 - L^*_2$$
$$\Delta a^* = a^*_1 - a^*_2$$
$$\Delta b^* = b^*_1 - b^*_2$$

The larger the ΔE^*-value, the greater the color difference becomes. It is known that a ΔE^*-value of 1.0 is just discernable by the average human eye [6].

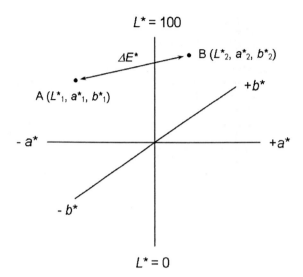

Fig. 2 Definition of the CIE 1976 $L^*a^*b^*$ (CIELAB) color difference, ΔE^*.

Materials and Methods

Sample Preparation

Table 1 gives analyzed chemical compositions of the alloys examined. The binary AP2, AP5, and AP10 alloys, containing about 2, 5, and 10 at.% Pt respectively, were prepared to investigate the effects of the addition of Pt to Au on its optical properties. The binary AP10 alloy was used as a reference to extract the effects of various alloying elements on the optical properties of Au-Pt-based high noble alloys. Further, the ternary AP10-In2 alloy was employed as a reference to evaluate the effects of the addition of the fourth element on the optical properties of the ternary Au-Pt-In alloy. To extract the effects of the third and fourth elements on the optical properties of the parent alloy, the ratio of Pt to Au contents in atomic percentage was controlled to be constant in all the ternary and quaternary alloys, as indicated in the right hand column in Table 1.

All the experimental alloys were prepared from high-purity constituent metals in a high-frequency induction furnace under an argon atmosphere. The ingot obtained was subjected to alternate cold rolling and homogenizing heat treatment at high temperatures. A number of homogenized plate samples sized 10 x 10 x 0.5 mm^3 were obtained.

Table 1 Chemical compositions of the alloys examined (analyzed values)

Alloy	Code*	Composition (at.%)						*Pt/Au*
		Au	Pt	In	Fe	Zn	Sn	
AP2	AP2	98.04	1.96	0	0	0	0	*0.020*
AP5	AP5	95.09	4.91	0	0	0	0	*0.052*
AP10	AP10	90.11	9.89	0	0	0	0	*0.110*
AP10-In1.0	In1.0	89.12	9.93	0.95	0	0	0	*0.111*
AP10-In1.7	In1.7	88.35	9.91	1.74	0	0	0	*0.112*
AP10-Fe0.8	Fe0.8	89.19	10.04	0	0.77	0	0	*0.113*
AP10-Fe1.9	Fe1.9	88.31	9.78	0	1.91	0	0	*0.111*
AP10-Zn1.7	Zn1.7	88.47	9.83	0	0	1.70	0	*0.111*
AP10-Sn0.9	Sn0.9	89.15	9.93	0	0	0	0.92	*0.111*
AP10-Sn2.0	Sn2.0	88.22	9.81	0	0	0	1.97	*0.111*
(AP10-In2)-Fe1.0	In2+Fe1.0	87.26	9.72	2.01	1.01	0	0	*0.111*
(AP10-In2)-Fe1.7	In2+Fe1.7	86.59	9.67	2.01	1.73	0	0	*0.112*
(AP10-In2)-Zn2.1	In2+Zn2.1	86.34	9.55	1.99	0	2.12	0	*0.111*
(AP10-In2)-Sn1.0	In2+Sn1.0	87.33	9.77	1.95	0	0	0.95	*0.112*

* The codes in this column were used in Figs. 10 and 11.

Spectrophotometric Colorimetry

Three plate samples from each alloy were individually embedded in cold-curing-type epoxy resin and subjected to a wet grinding phase. The wet grinding of the plate samples was performed using waterproof SiC abrasive papers adhered to a turntable in an automatic polishing apparatus (MA-150, Musashino Denshi Co., Ltd., Tokyo, Japan). After grinding down to 2000-grit finish, the samples were loaded on a polishing cloth adhered to a turntable in an automatic polishing apparatus (Doctor-Lap ML-180, Maruto Instrument Co., Ltd., Tokyo, Japan) and then successively polished using alumina suspension with a grain diameter of 0.3 μm and 0.06 μm.

After rinsing with pure water and drying, the polished samples were mounted on a computer-controlled spectrophotometer (CM-3600d, Konica Minolta Sensing, Inc., Osaka, Japan), and the spectral reflectance data for the incident CIE (Commission Internationale de l'Eclairage) standard illuminant D65 were collected at 10 nm intervals from 360 to 740 nm under the visual field of 10°. Three-dimensional color coordinates, *i.e.*, $L*$ (lightness), $a*$ (red-green chromaticity index), $b*$ (yellow-blue chromaticity index) in the CIELAB color space, were determined for each sample. The standard illuminant D65 is the simulated natural daylight with a 6504 K blackbody temperature. Due to the highly reflective nature of the sample surface, the specular component included (SCI) data were collected in the present color measurements. The spectral reflectance curves and three-dimensional CIELAB color coordinates for pure Au and pure Pt were also obtained for comparison. Three polished samples from each alloy and from pure metals were subjected to the measurements.

Results

Spectral Reflectance Curves

Binary Au-Pt Alloys
Fig. 3 shows spectral reflectance curves for Au, Pt, and the binary Au-Pt alloys. In this figure and the following similar figures showing spectral reflectance curves, average reflectance value and standard deviations for three measurements at each wavelength were plotted against wavelength. Because standard deviations for three reflectance data at each wavelength were so small, error bars indicating standard deviations were embedded in the data points. Reflectance from pure Au was very high in the long-wavelength range but very low in the short-wavelength range. As a result, a pronounced step was observed near 515 nm in the spectral reflectance curve. To the contrary, reflectance from pure Pt was considerably low in the long-wavelength range and slightly decreased with decreasing wavelength. Consequently, a fairly flat spectral reflectance curve was observed for pure Pt.

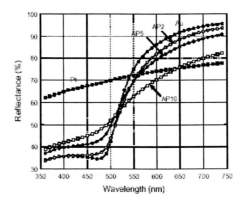

Fig. 3 Spectral reflectance curves for Au, Pt, and the binary Au-Pt alloys.

The alloying addition of a small amount of Pt to Au caused a drastic change in the shape of the spectral reflectance curve. That is, reflectance in the long-wavelength range markedly decreased and reflectance in the short-wavelength range slightly increased with the addition of Pt to Au. As a result, the steep step near 515 nm in the spectral reflectance curve for pure Au became less pronounced.

Ternary AP10-X Alloys

The effects of the alloying addition of the third element to the binary AP10 alloy on its spectral reflectance curve were investigated. Fig. 4 demonstrates spectral reflectance curves for the binary AP10 and the ternary AP10-Fe0.8 and AP10-Fe1.9 alloys. It is clear that the alloying addition of a small amount of Fe markedly increased reflectance in the wavelengths longer than about 500 nm. Degree of increases in reflectance in the long-wavelength range apparently increased with increasing Fe content. As a result, the step of the spectral reflectance curve near 515 nm for the samples apparently became pronounced with increasing Fe content. Similar effects of Sn addition on the spectral reflectance curves were evidenced when the parent AP10 alloy was alloyed with a small amount of Sn, as shown in Fig. 5.

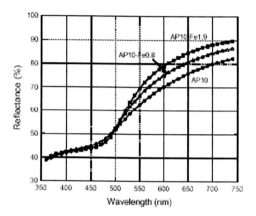

Fig. 4 Spectral reflectance curves for the binary AP10 and the ternary AP10-Fe0.8 and AP10-Fe1.9 alloys.

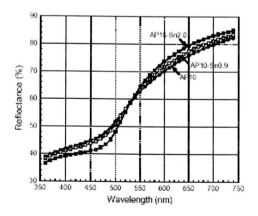

Fig. 5 Spectral reflectance curves for the binary AP10 and the ternary AP10-Sn0.9 and AP10-Sn2.0 alloys. (This figure was cited from our previous paper [13] with official permission by the publisher.)

Fig. 6 shows spectral reflectance curves for the parent AP10 and AP10-Zn1.7 alloys. It is noted that although the amount of Zn added was as much as 1.7 at.%, the effects of Zn addition on the spectral reflectance curve was very weak compared to those of Fe and Sn.

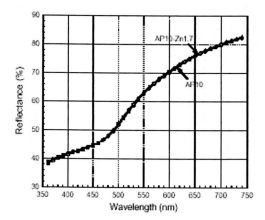

Fig. 6 Spectral reflectance curves for the binary AP10 and the ternary AP10-Zn1.7 alloys. (This figure was cited from our previous paper [13] with official permission by the publisher.)

Quaternary (AP10-In2)-X Alloys

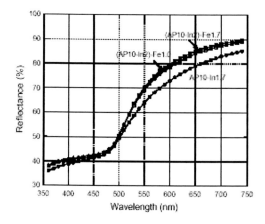

Fig. 7 Spectral reflectance curves for the ternary AP10-In1.7 and the quaternary (AP10-In2)-Fe1.0 and (AP10-In2)-Fe1.7 alloys.

In this subsection, the effects of the alloying addition of the fourth element to the ternary AP10-In2 alloy will be presented. Fig. 7 shows spectral reflectance curves for the ternary AP10-In1.7 and the quaternary (AP10-In2)-Fe1.0 and (AP10-In2)-Fe1.7 alloys. The alloying addition of Fe to the parent AP10-In2 alloy effectively increased reflectance in the wavelengths longer than about 500 nm. Again, the step near 515 nm in the spectral reflectance curve became pronounced with the addition of Fe. Similar effects were also evidenced when the parent AP10-In2 alloy was alloyed with a small amount of Sn, as shown in Fig. 8. On the other hand, the effects of Zn addition on the spectral reflectance curve were very weak as in the case of the Zn addition to the AP10 alloy (See Fig. 6).

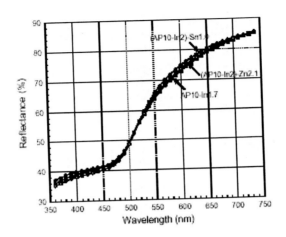

Fig. 8 Spectral reflectance curves for the ternary AP10-In1.7 and the quaternary (AP10-In2)-Sn1.0 and (AP10-In2)-Zn2.1 alloys. (This figure was cited from our previous paper [13] with official permission by the publisher.)

The CIELAB Color Coordinates

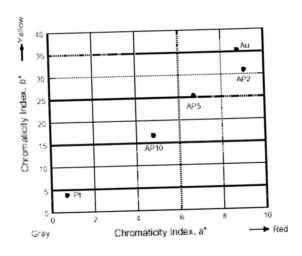

Fig. 9 Chromaticity indices, a^* and b^*, evaluated for Au, Pt, and the binary Au-Pt alloys.

Fig. 9 shows chromaticity indices, a^* and b^*, in the three-dimensional CIELAB color space for Au, Pt, and the binary Au-Pt alloys. It is clearly shown that the alloying addition of Pt to Au considerably decreased a^*-value and markedly decreased b^*-value, indicating the strong decolorizing effect of Pt. It is noted that the addition of about 10 at.% Pt to Au decreased both indices by more than half the distance between Au and Pt.

The effects of alloying additions of various base metals to the binary AP10 alloy on its chromaticity indices are summarized in Fig. 10. The sole addition of In, Fe, or Sn to the parent AP10 alloy considerably increased both indices. As a result, the data points for the ternary alloys containing these third elements apparently moved towards the point for pure Au. This means that the alloying addition of these base metals to the AP10 alloy gives a gold tinge to the parent Au-Pt alloy. The dual addition of In and Fe/Sn to the parent AP10 alloy further increased both $a*$ and $b*$ values, indicating the increased effect of giving a gold tinge to the parent AP10 alloy. On the other hand, the effects of the addition of Zn to the AP10 alloy on its chromaticity indices were very weak.

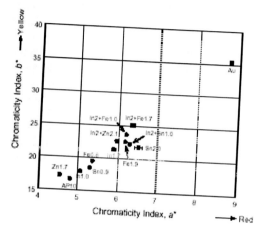

Fig. 10 Chromaticity indices, $a*$ and $b*$, evaluated for all the ternary and quaternary alloys examined. The chromaticity indices for pure Au and the binary AP10 alloy were also displayed for comparison.

The effects of alloying additions of various base metals to the binary AP10 alloy on its lightness index $L*$ are demonstrated in Fig. 11. The addition of In, Sn, and Zn to the binary AP10 alloy slightly raised the $L*$-value. On the other hand, a pronounced increase in lightness $L*$-value was observed when the parent AP10 alloy was alloyed with Fe.

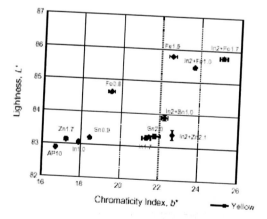

Fig. 11 Lightness $L*$, $vs.$ chromaticity index $b*$ plot for all the ternary and quaternary alloys examined. The data for the binary AP10 alloy was also displayed for comparison.

The CIELAB Color Difference, ΔE^*

The effects of various alloying elements on the color difference, ΔE^*, between the binary AP10 and the ternary alloys containing Fe, Sn, In, or Zn, are presented in Fig. 12. It is shown that the involvement of Fe causes a prominent color change, while the addition of Zn causes a very small change in color. Since the ΔE^*-value of 1.0 is just discernible [6], color of the AP10-Zn1.7 alloy may be scarcely distinguished from that of the parent AP10 alloy, although the amount of Zn added is as much as about 1.7 at.%.

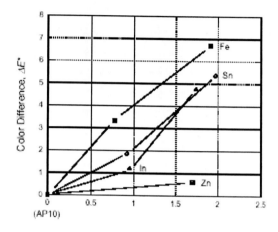

Fig. 12 The effects of the various alloying elements on the CIELAB color difference parameter, ΔE^*, between the ternary alloys and the binary AP10 alloy.

Discussion

Decolorizing Mechanism of Platinum Addition

The perceived color of a metal is determined by the wavelength distribution of the radiation that is reflected and not absorbed [11]. For example, both Au and Cu have rather low reflectance at the short-wavelength range of the visible spectrum and so yellow and red will consequently be reflected to a greater degree [12]. These characteristic distributions of spectral reflectance for Au and Cu determine their specific colors. Therefore, both the position of the absorption edge of the spectral reflectance curve and the slope of the spectral reflectance curve near the absorption edge are considered to greatly affect the perceived color of metals and alloys [13].

Before discussing the decolorizing mechanism of Pt addition, we will begin with characterization of the spectral reflectance curve for pure Au. To find the position of the absorption edge and slope of the spectral reflectance curve at its absorption edge, differential coefficient of reflectance, dR/dL, at each wavelength was calculated from the reflectance data for Au. That is, dR/dL value at each wavelength was obtained by differentiating spectral reflectance data (R) with respect to wavelength (L). By plotting these dR/dL-values against wavelength, we obtain a dR/dL-curve shown in Fig. 13. The line with solid circles is the

spectral reflectance curve for pure Au, and the line with open circles is the corresponding
dR/dL-curve. The peak position of the dR/dL-curve, indicated by a double arrow, corresponds
to the absorption edge of the spectral reflectance curve and the peak height corresponds to the
slope of the spectral reflectance curve at its absorption edge. It is shown that the absorption
edge for Au is located at about 515 nm (approximately 2.4 eV). This energy is very close to
the previously reported values of 2.45 eV [14] and approximately 2.3 eV [15, 16]. This
energy is interpreted to correspond to the energy required for transition of d-electrons to the
energetically higher conduction band at the Fermi level [14, 16, 17].

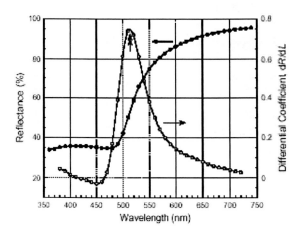

Fig. 13 Spectral reflectance curve (solid circles) and its differential coefficient curve (open circles) for
pure Au. (This figure was cited from our previous paper [13] with official permission by the publisher.)

Based on this characterization of the spectral reflectance curve for Au, the decolorizing
mechanism of Pt addition will be discussed. Fig. 14 shows wavelength distributions of
differential coefficient, dR/dL, for Au, Pt, and the binary Au-Pt alloys. These curves were
obtained from reflectance data presented in Fig. 3. The peak positions of the dR/dL-curves for
the binary Au-Pt alloys, indicated by double arrows, were not affected by the alloying
addition of Pt within the limit of experimental error. On the other hand, the peak height
markedly decreased with increasing Pt content. These facts indicate that the step in the
spectral reflectance curve near 515 nm for the binary Au-Pt alloys became less pronounced
with increasing Pt content. By reviewing the spectral reflectance curves for the Au-Pt alloys
in Fig. 3, it is clear that the considerable decrease in the slope of the spectral reflectance curve
at its absorption edge in the binary Au-Pt alloys is mainly caused by the significant decrease
in reflectance in the long-wavelength range in the visible spectrum.

Similar flattening phenomenon of the spectral reflectance curve was previously reported
in Au-Ni and Au-Pd systems [16]. Saeger and Rodies [16] found that when Au was alloyed
with Ni or Pd, reflectance in the long-wavelength range systematically decreased with
increasing Ni or Pd content. As a result, the step in reflectivity curve, which is responsible for
the yellow color of fine Au, became less pronounced [16]. They explained this decolorizing
effect of Ni or Pd in terms of "virtual bound states" first suggested by Friedel [18]. That is,
when monovalent noble metals are alloyed with transition metals like Ni or Pd, so-called
"virtual bound states" will occur [16]. In fact, Abelès [19] showed that the maxima of the
relatively broad absorption peaks caused by the virtual bound states lie at energies of about

0.8 eV (about 1550 nm) for Au-Ni and 1.7 eV (about 730 nm) for Au-Pd systems [16, 20]. This means that the reflectance in the low energy range (long-wavelength range) in the visible spectrum is decreased with the addition of Ni or Pd to Au. Accordingly, it is suggested that the "virtual bound states" may also occur when Au is alloyed with Pt like in the cases of Au-Ni and Au-Pd systems. This interpretation well explains the present experimental results that the reflectance in the long-wavelength range of the binary Au-Pt alloys systematically decreased with increasing Pt content.

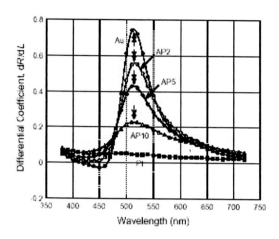

Fig. 14 dR/dL-curves for Au, Pt, and the binary Au-Pt alloys.

Factors Affecting the Optical Properties of Au-Pt-Based High Noble Alloys

To understand mechanism of variations of color with composition in the Au-Pt-based multi-component alloys, the effects of the alloying addition of the third element to the binary AP10 alloy on the position of the absorption edge and the slope of the spectral reflectance curve at its absorption edge were investigated. Fig. 15 shows wavelength distributions of differential coefficient, dR/dL, for the binary AP10 and the ternary AP10-Fe0.8 and AP10-Fe1.9 alloys. It was evidenced that the peak positions of the dR/dL-curves, indicated by double arrows, were not affected by the alloying addition of Fe. The peak height, on the other hand, significantly increased with increasing Fe content. Therefore, it was revealed that the alloying addition of Fe to the binary AP10 alloy did not move the position of the absorption edge but effectively increased the slope of the spectral reflectance curve at its absorption edge. Consequently, it is drawn that an increased gold tinge in the Fe-added Au-Pt alloys, shown in Fig. 10, is attributed to the increased slope of the spectral reflectance curve at its absorption edge. This increase in the slope was mainly caused by the pronounced increase in reflectance in the yellow to red range of the visible spectrum, as previously presented in Fig. 4. Similar analysis of the spectral reflectance data was also performed for the AP10-In and AP10-Sn alloys. It was revealed that the addition of In or Sn to the parent AP10 alloy did not change the position of the absorption edge but increased the slope of the spectral reflectance curve at its absorption edge [13]. Based on these experimental evidences, it can be mentioned that a gold tinge given to the ternary AP10-X (X = Fe, In, or Sn) alloys was due to the marked increase in slope of the spectral reflectance curve at its absorption edge, and not to the shift of the absorption edge.

The causes of a gold tinge given to the Fe-added AP10-In2 alloys were investigated in a similar manner. Fig. 16 shows wavelength distributions of the dR/dL-values for the ternary AP10-In1.7 alloy and the quaternary (AP10-In2)-Fe1.0 and (AP10-In2)-Fe1.7 alloys obtained from the spectral reflectance data presented in Fig. 7. It is seen that the addition of Fe to the parent AP10-In2 alloy did not move the position of the absorption edge but apparently increased the slope, dR/dL, of the spectral reflectance curve at its absorption edge. It is clear that the increase in slope of the spectral reflectance curve at its absorption edge, not the shift of the absorption edge, was responsible for giving a gold tinge to the parent Au-Pt-In alloy. Fig. 17 summarizes the relationships between chromaticity indices, a^* and b^*, and the slope, dR/dL, of the spectral reflectance curve at its absorption edge for all the alloys examined except for the Au-rich binary AP2 and AP5 alloys. It is clearly shown that both chromaticity indices systematically increased with increasing dR/dL-value at the absorption edge.

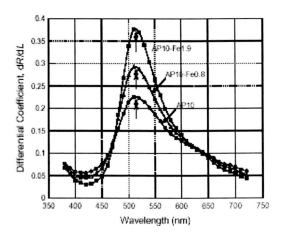

Fig. 15 dR/dL-curves for the binary AP10 and the ternary AP10-Fe0.8 and AP10-Fe1.9 alloys.

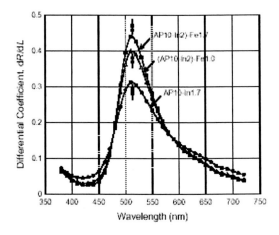

Fig. 16 dR/dL-curves for the ternary AP10-In1.7 and the quaternary (AP10-In2)-Fe1.0 and (AP10-In2)-Fe1.7 alloys.

The above-described analysis of the spectral reflectance curves for the ternary AP10-X and quaternary (AP10-In2)-X alloys leads to the expression that the alloying additions of small amounts of the third or fourth elements to the parent AP10 or AP10-In2 alloys, respectively, did not move the position of the absorption edge near 515 nm but increased the slope of the spectral reflectance curve at the absorption edge depending on the alloying element and its content. It was evidenced that with increasing slope of the spectral reflectance curve at its absorption edge, both chromaticity indices increased, giving a gold tinge to the parent AP10 or AP10-In2 alloys.

Since optical properties of metals and alloys are determined by the interaction between the photons of incident light and the electrons in a metallic material, it is suggested that the number of valence electrons per atom, i.e., electron:atom ratio (e/a), in the metallic material may greatly affect both absorption and reflection processes [13]. That is, degrees of both reflection and absorption may increase with increasing e/a-values. In other words, the slope of the spectral reflectance curve at its absorption edge may increase with increasing e/a-values. To examine this hypothesis, relationship between e/a-values in the experimental alloys and dR/dL-values at the absorption edge was investigated.

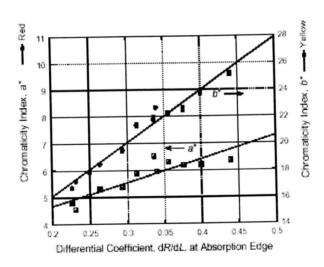

Fig. 17 The relationships between chromaticity indices, a* and b*, in the CIELAB color space and the slope of the spectral reflectance curve at its absorption edge for all the alloys examined except for the Au-rich binary AP2 and AP5 alloys.

An electron:atom ratio, e/a, of an alloy was estimated according to the following equation:

$$e/a = \Sigma f_i N_i \tag{2}$$

where, f_i is the mole fraction of the component i in the multi-component alloy, and N_i is the number of valence electrons per atom for the component i [13]. According to the previous studies, the number of valence electrons per atom is 1 for Au [21, 22], 0 for Pt [21, 23], 2 for Zn [21, 24, 25], 3 for In [21, 22, 24], 4 for Sn [21, 25]. To determine the number of valence electrons per atom for Fe, the Moessbauer spectroscopy was performed on the present AP10-

Fe1.9 alloy [26]. According to such Moessbauer studies, the isomer shift of the Fe atoms in the present AP10-Fe1.9 alloy was similar to metallic Fe [26]. Therefore, the number of valence electrons per atom for Fe was tentatively assigned to 4, following a similar argument as for the other alloys [26]. Based on these reported numbers of valence electrons per atom for the constituent elements and chemical compositions listed in Table 1, e/a-values for the present Au-Pt-based multi-component alloys were estimated according to the equation (2). The results of such estimation were summarized in Table 2.

Fig. 18 shows the relationship between the slope, dR/dL, of the spectral reflectance curve at its absorption edge and number of valence electrons per atom, e/a, for all the alloys listed in Table 2. There is an apparent positive relationship between these parameters. That is, the increase in e/a-value effectively raises the slope of the spectral reflectance curve at its absorption edge. By combining relationships presented in Figs. 17 and 18, we are able to know the effects of e/a-value in the alloy on chromaticity indices, $a*$ and $b*$. Fig. 19 shows the relationships between chromaticity indices, $a*$ and $b*$, in the CIELAB color space and electron:atom ratio for all the alloys examined. It is clear that with increasing e/a-value the red-green chromaticity index $a*$ slightly increased and the yellow-blue chromaticity index $b*$ markedly increased. This means that with increasing e/a-value a gold tinge in a Au-Pt-based multi-component alloy is increased. Therefore, it is concluded that the number of valence electrons per atom (e/a) in an alloy is a controlling factor for chromaticity indices, $a*$ and $b*$, of Au-Pt-based high noble alloys and that a greater degree of gold tinge is given to the alloy with increasing e/a-value.

Table 2 Estimated electron:atom ratio (e/a) of the alloys.

Alloy	Code*	e/a
AP10	AP10	0.9011
AP10-In1.0	In1.0	0.9197
AP10-In1.7	In1.7	0.9357
AP10-Fe0.8	Fe0.8	0.9227
AP10-Fe1.9	Fe1.9	0.9595
AP10-Zn1.7	Zn1.7	0.9187
AP10-Sn0.9	Sn0.9	0.9283
AP10-Sn2.0	Sn2.0	0.9610
(AP10-In2)-Fe1.0	In2+Fe1.0	0.9733
(AP10-In2)-Fe1.7	In2+Fe1.7	0.9954
(AP10-In2)-Zn2.1	In2+Zn2.1	0.9655
(AP10-In2)-Sn1.0	In2+Sn1.0	0.9698

*The codes in this column were used in Figs. 10 and 11.

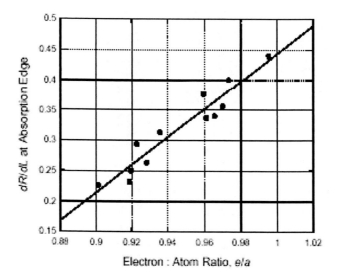

Fig. 18 The relationship between the slope of the spectral reflectance curve at its absorption edge and the electron:atom ratio in the alloy for all the alloys except for the Au-rich binary AP2 and AP5 alloys.

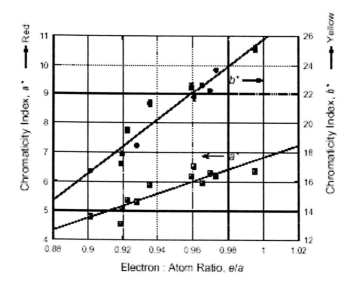

Fig. 19 The relationships between chromaticity indices, a^* and b^*, in the CIELAB color space and electron:atom ratio for all the alloys examined except for the Au-rich binary AP2 and AP5 alloys.

Based on the "virtual bound states" theory explained in the preceding subsection, it is suggested that the alloying addition of a base metal with a high number of valence electrons per atom may weaken the degree of virtual bound states in the Au-Pt-based alloys. As a result, reflectance in the long-wavelength range in the visible spectrum may increase with increasing e/a-value in an alloy. This may be the reason that chromaticity indices, a^* and b^*, in the CIELAB color space increased with increasing e/a-value in the Au-Pt-based high noble alloys.

Conclusions

Optical properties of the binary Au-Pt alloys and of the Au-Pt-based high noble alloys containing small amounts of various base metals with different numbers of valence electrons were investigated by means of spectrophotometric colorimetry. Careful analysis of the collected spectral reflectance curves led to the following conclusions.

The alloying addition of Pt to Au markedly decolorized. This decolorizing effect of Pt addition was found to be due to the flattening of the spectral reflectance curves in the visible spectrum. This flattening of the spectral reflectance curve, occurred in the binary Au-Pt alloys, was caused by the marked decrease in reflectance in the long-wavelength range and the slight increase in reflectance in the short-wavelength range. It was suggested that this flattening phenomenon of the spectral reflectance curve in the Au-Pt system was due to the possible formation of the "virtual bound states" as in the cases of Au-Pd and Au-Ni systems.

The alloying addition of small amounts of base metals with a high number of valence electrons to the parent Au-Pt alloy effectively increased a gold tinge. This pronounced effects of base metals on the color of the Au-Pt-based high-noble alloys was controlled by the number of valence electrons per atom, e/a, in an alloy. That is, with increasing e/a-value in an alloy, the slope of the spectral reflectance curve at its absorption edge near 515 nm (approximately 2.4 eV) effectively increased. As a result, a red-green chromaticity index a^*-value slightly increased and a yellow-blue chromaticity index b^*-value markedly increased with increasing e/a-value, giving an increased gold tinge to the parent alloy.

The above-described mechanism of color change with chemical composition in Au-Pt-based multi-component alloys is expected to be useful in controlling the color of the Au-Pt-based high noble alloys.

Acknowledgements

This research was supported in part by The Grant-in-Aid for Scientific Research from Nagasaki University and from Nagasaki Advanced Technology Development Council. The author is grateful to Ishifuku Metal Industry Co., Ltd., (Tokyo, Japan) for the preparation of sample alloys.

References

[1] R. M. German: *Gold Bulletin*, **13**(2), 57-62, 1980.

[2] H. Knosp, R. J. Holliday, C. W. Corti: *Gold Bulletin*, **36**(3), 93-102, 2003.

[3] J.-M. Meyer: Porcelain-Metal Bonding in Dentistry, in "*Concise Encyclopedia of Medical & Dental Materials*", Edited by D. Williams, Pergamon Press, Oxford, pp. 307-312, 1990.

[4] R. van Noort: "*Introduction to Dental Materials*", Mosby, pp. 215-224, 1994.

[5] *Platinum 2004*, p. 11, Johnson Matthey, London, England, May 2004.

[6] R. M. German, M. M. Guzowski, D. C. Wright: *Journal of Metals*, **32**(3), 20-27, 1980.

[7] R. M. German: *International Metals Reviews*, **27**(5), 260-288, 1982.

[8] J. Leuser: *Metall*, **3**(7/8), 105-110, 1949.

[9]   R. A. Fuys, C. W. Fairhurst, W. J. O'Brien: *Journal of Biomedical Materials Research*, **7**, 471-480, 1973.

[10] R. M. German: *Journal of Dental Research*, **59**(11), 1960-1965, 1980.

[11] W. D. Callister, Jr.: *"Materials Science and Engineering: An Introduction"*, John Wiley & Sons, Inc., New York, pp. 535-536, 1985.

[12] R. J. D. Tilley: *"Colour and Optical Properties of Materials"*, John Wiley & Sons, Ltd., Chichester, England, pp. 232-234, 2000.

[13] T. Shiraishi, Y. Takuma, E. Miura, Y. Tanaka, K. Hisatsune: *Journal of Materials Science: Materials in Medicine*, **14**(12), 1021-1026, 2003.

[14] M. L. Thèye: *Physical Review B*, **2**(8), 3060-3078, 1970.

[15] H. Fukutani and O Sueoka: Optical Properties of Ag-Au Alloys, in *"Optical Properties and Electronic Structure of Metals and Alloys"*, Edited by F. Abelès, North-Holland Publishing Company, Amsterdam, pp. 565-573, 1966.

[16] K. E. Saeger and J. Rodies: *Gold Bulletin*, **10**, 10-14, 1977.

[17] B. R. Cooper, H. Ehrenreich, and H. R. Philipp: *Physical Review*, **138**(2A), 494-507, 1965.

[18] J. Friedel: *Canadian Journal of Physics*, **34**, 1190-1211, 1956.

[19] F. Abelès: Proprietes Optiques des Alliages Metalliques, in *"Optical Properties and Electronic Structure of Metals and Alloys"*, Edited by F. Abelès, North Holland Publishing Company, Amsterdam, pp. 553-564, 1966.

[20] T. Shiraishi, K. Hisatsune, Y. Tanaka, E. Miura and Y. Takuma: *Gold Bulletin*, **34**(4), 129-133, 2001.

[21] N. F. Mott and H. Jones: *"The Theory of the Properties of Metals and Alloys"*, Dover Publications, Inc., New York, pp. 172-173, 1958.

[22] H. Sato and R. S. Toth: *Physical Review*, **124**, 1833-1847, 1961.

[23] K. Ohshima and D. Watanabe: *Acta Crystallographica A*, **29**, 520-526, 1973.

[24] C. Kittel: *"Introduction to Solid State Physics"*, 2nd Edition, John Wiley & Sons, Inc., New York, p. 320, 1956.

[25] R. E. Reed-Hill, *"Physical Metallurgy Principles"*, 2nd Edition, PWS-KENT Publishing Company, Boston, pp. 112-115, 1973.

[26] J. Linden (Department of Physics, Abo Akademi, Turku, Finland), *Private communication*

In: Trends in Materials Science Research
Editor: B.M. Caruta, pp. 101-115

ISBN: 1-59454-367-4
© 2006 Nova Science Publishers, Inc.

Chapter 4

YIELD SURFACE OF SHAPE MEMORY ALLOYS

W.M. Huang[1] and X.Y. Gao[2]

School of Mechanical and Aerospace Engineering, Nanyang Technological University,
50 Nanyang Avenue, Republic of Singapore

Abstract

The paper investigates three issues about the yield surface (transformation start stress) of shape memory alloys (SMAs). The first is the two-pole phenomenon. We show that this phenomenon can happen in a SMA if there is a variation in the Young's modulus upon the phase transformation. The second is about multiphase transformation in some SMAs. In such a case, the yield surface may be determined by the superposition method. We compare our predictions with the experimental results of two SMAs. The last one is the difference in the yield surface between the phase transformation and martensite reorientation. It reveals that despite that there is not much difference in the yield surface in the $(\sigma_1-\sigma_2,\ \sigma_3=0)$ plane between the phase transformation and martensite reorientation, the yield surface of the phase transformation is a slant cone if there is a volume variation in the phase transformation, while that of the martensite reorientation is always a slant column as no volume change is associated in the martensite reorientation.

Keywords: Shape memory alloys, Yield surface, Martensitic transformation, Reorientation

Introduction

Due to their incomparably huge recoverable strain (up to 10%) and actuation stress (400 MPa or more), shape memory alloys (SMAs) attract a lot of attention in recent years (e.g., Funakubo 1987, Otsuka and Wayman 1998, Huang 2002). The unique shape memory behavior in SMAs is resulted by two types of transformations at the crystal lattice level, namely, phase transformation between austenite and martensite, and martensite reorientation. As illustrated in Figure 1, there are normally three phases in a SMA, i.e., austenite, twinned

---

[1] E-mail address: mwmhuang@ntu.edu.sg, Tel: (0065) 67904859, Fax: (0065) 67911859,
[2] Currently with the Faculty of Engineering, National University of Singapore, Singapore.

martensite and detwinned martensite. Six types of transformations among these three phases are possible, namely (Huang and Zhang 2003)

1   Austenite transforms into detwinned martensite upon loading ($A \Rightarrow DM$);
2   Detwinned martensite transforms back to austenite upon unloading at a high temperature or upon heating ($DM \Rightarrow A$);
3   Detwinned martensite (variant k) transforms into another detwinned martensite (variant l) upon loading ($DM_k \Rightarrow DM_l$);
4   Twinned martensite transforms into detwinned martensite upon loading ($TM \Rightarrow DM$);
5   Austenite transforms into twinned martensite upon cooling ($A \Rightarrow TM$);
6   Twinned martensite transforms into austenite upon heating ($TM \Rightarrow A$).

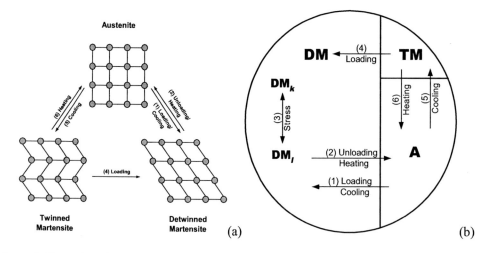

Figure 1 Illustration of transformations/phases in SMAs. (a) Phase variation upon transformation; (b) three phases and six transformations. (Huang and Zhang 2003)

At present, the lattice structures of the austenite and martensite phases of most SMAs are well known, despite some occasional contradictions in the literature. However, the application of SMAs is still lagged by the lack of convenient and reliable computation tools for simulation and analysis of SMAs. Most of the thermomechanical models proposed in the literature either have tremendous difficulty in describing the macro-thermomechanical behavior of SMAs with good accuracy or are too complex to be applied in a real engineering application (e.g. Huang 1999b, Zhu et al 2002). It turns out that a simple but reliable computation algorithm is still highly demanded at present.

One of such algorithms is the one proposed by Huang and Zhu (2002), which can well cope the situation that a SMA is under proportional loading. Similar to that in the traditional theory of plasticity, one of the pre-requirements is the yield surface (in SMAs, it is the transformation start stress instead of yielding). Huang (1999a) proposed a simple approach to determine the yield surface of SMAs. It is based on the lattice structures of the austenite and martensite phases and the correspondence variants. Following this approach, in Gao and Huang (2002), the yield surfaces of four non-textured polycrystalline SMAs, namely, NiTi, NiAl, CuAlNi and CuZnGa, were obtained. In addition, the effects of different Young's moduli of austenite and martensite were investigated. Apart from these, it also explained the reason behind the V-shape phenomenon in the transformation start stress verse temperature relationship. According to this study, the V-shape phenomenon should only occur in the cooling process of SMAs. Huang (2004) further demonstrated that the difference in the Young's modulus causes the non-linearity in the phase transformation start stress verse temperature relationship in SMAs.

The purpose of this study is to discuss three important issues about the yield surface of non-textured polycrystalline SMAs. After a brief summary of the framework, which is mainly based on the previous works presented in Huang (1999a), Gao and Huang (2002), and Huang and Zhu (2002), we investigate the two-pole phenomenon in the three-dimensional yield surface of SMAs (in the principal stress space). Next, we study the case of multiphase transformation in some SMAs. Subsequently, we discusse the difference in yield surface upon the phase transformation and martensite reorientation. Conclusions are summarized in the end of this chapter.

Framework

The theoretical framework proposed in Huang (1999a) is based on the lattice structures of the austenite and martensite phases and the transformation relationships of SMAs. Similar to the slip systems in traditional crystallographic plasticity, there might be m transformation systems in a SMA single crystal upon loading. Provided that the difference in Young's moduli of austenite and martensite is small, the criterion for the start of phase transformation in this single crystal can be expressed as

$$\mathop{\mathrm{Max}}_{1 \le l \le m}\left(\Sigma : \varepsilon_l^p\right) = K \tag{1}$$

where K, driving energy, is a constant at a given temperature. Σ and ε_l^p $\left(l = 1,2,\ldots m\right)$ are the applied stress and strain associated with the transformation, respectively.

Given a stress state, Eqn. (1) reveals that the real transformation strain ε_l^p of a grain corresponds to the minimum value of the critical stress, i.e., the lowest transformation start stress. This pair of transformation strain and stress is most likely to happen. Consider the phase transformation as a special case of yielding, provided that the lattice structures of martensite and austenite of a SMA are known, the yield surface of the SMA single crystal can be obtained from Eqn. (1).

For convenience, as a traditional practice, in the following discussion on polycrystalline SMAs, we shall assume that all the yield stress to be normalized by the yield stress of uni-axial tension.

Given a polycrystalline SMA with n grains, the transformation strain of the grain k ($k = 1, 2, \ldots n$) with respect to the global coordinate system is expressed as

$$\mathbf{E}_{kl}^p = \mathbf{R}_k^T \varepsilon_l^p \mathbf{R}_k; \quad \left(k = 1,2,\ldots n\right) \tag{2}$$

Here, $\mathbf{R}_k$ is the rotation matrix (from the local coordinate system, which is attached to the austenite lattice structure of the grain k, to the global coordinate system).

To extend Eqn. (1) into non-textured polycrystalline SMAs, two schemes were proposed by Huang and Zhu (2002).

Given a stress state Σ^0 (in any magnitude except 0), the first scheme yields that the *exact* yield stress Σ_{Max}^p (relative to that of uni-axial tension) is

$$\Sigma_{\mathrm{Max}}^p = \frac{\Sigma^0}{\alpha_{\mathrm{Max}}} \tag{3}$$

where

$$\alpha_{\mathrm{Max}} = \frac{\underset{1 \leq k \leq n}{\mathrm{Max}}\left(K_{\mathrm{Max}}^{k,0}\right)}{\underset{1 \leq k \leq n}{\mathrm{Max}}\left(K_{\mathrm{Max}}^{k,\,\mathrm{tension}}\right)} \tag{4}$$

and

$$K_{\mathrm{Max}}^{k,0} = \underset{1 \leq l \leq m}{\mathrm{Max}}(\Sigma^0 : \mathbf{E}_{kl}^p) \tag{5}$$

$$K_{\mathrm{Max}}^{k,\,\mathrm{tension}} = \underset{1 \leq l \leq m}{\mathrm{Max}}(\Sigma^{\mathrm{tension}} : \mathbf{E}_{kl}^p) \tag{6}$$

Here, $\Sigma^{\mathrm{tension}}$ stands for the stress state of uni-axial tension. This is the maximum scheme since the start of yielding depends only on one grain that produces the maximum K. Interactions among grains are ignored. This is analogous to have all grains in series, one of the traditional approaches adopted in the classic theory of plasticity.

In the second scheme, under a given applied stress, the interactions among grains are counted by means of a simple average of the driving energy over all grains, i.e.,

$$\Sigma_{\mathrm{Ave}}^p = \frac{\Sigma^0}{\alpha_{\mathrm{Ave}}} \tag{7}$$

where

$$\alpha_{\mathrm{Ave}} = \frac{\sum_{k=1}^{n} K_{\mathrm{Max}}^{k,0}}{\sum_{k=1}^{n} K_{\mathrm{Max}}^{k,\,\mathrm{tension}}} \tag{8}$$

This scheme is analogous to let all grains be in parallel, another common approach adopted in the traditional theory of plasticity. As it is the driving energies of all grains that are averaged, this scheme is different from many traditional methods, such as, Voigt approximation, Reuss approximation and Taylor's method (Mura 1987, Taylor 1938). The advantage of not using any inclusion theory in this framework is the simplicity that otherwise will disappear.

Note that if $\mathbf{R}_k$ is expressed by three Euler angles (φ_1, ϕ and φ_2), the orientation element dg is given by the following expression (Bunge 1982),

$$dg = \frac{1}{8\pi^2}\sin\phi \, d\varphi_1 d\phi d\varphi_2 \tag{9}$$

Thus, Eqn. (8) may be rewritten as

$$\alpha_{Ave} = \frac{\int_0^{2\pi}\int_0^{\pi}\int_0^{2\pi} K_{Max}^{k,0} \sin\phi \, d\varphi_1 d\phi d\varphi_2}{\int_0^{2\pi}\int_0^{\pi}\int_0^{2\pi} K_{Max}^{k,\,tension} \sin\phi \, d\varphi_1 d\phi d\varphi_2} \tag{10}$$

It is apparent that the real yield surface might be somewhere close to the average scheme, since the interactions among grains in SMAs should be strong, while the maximum scheme is the extreme situation that gives one boundary.

We can prove that the maximum scheme is equivalent to the following expression

$$\varepsilon_1^p \sigma_1 + \varepsilon_2^p \sigma_2 + \varepsilon_3^p \sigma_3 = K \tag{11}$$

where σ_1, σ_2 and σ_3 $(\sigma_1 \geq \sigma_2 \geq \sigma_3)$ are the principal stress, and ε_1^p, ε_2^p and ε_3^p are the principal strain (eigenvalue) of ε_I^p ($\varepsilon_1^p \geq \varepsilon_2^p \geq \varepsilon_3^p$, $\varepsilon_1^p \geq 0$, $\varepsilon_3^p \leq 0$).

Furthermore, the parametric solution has been obtained to describe the yield surface in the three-dimensional principal stress space as (Huang and Zhu 2002)

$$\begin{cases} \sigma_1 = \sigma_0 + \dfrac{\left(C - 3\varepsilon_0^p \sigma_0\right)}{(2-\beta)\varepsilon_1^p + (2\beta-1)\varepsilon_2^p - (1+\beta)\varepsilon_3^p}(2-\beta) \\[4mm] \sigma_2 = \sigma_0 + \dfrac{\left(C - 3\varepsilon_0^p \sigma_0\right)}{(2-\beta)\varepsilon_1^p + (2\beta-1)\varepsilon_2^p - (1+\beta)\varepsilon_3^p}(2\beta-1) \\[4mm] \sigma_3 = \sigma_0 - \dfrac{\left(C - 3\varepsilon_0^p \sigma_0\right)}{(2-\beta)\varepsilon_1^p + (2\beta-1)\varepsilon_2^p - (1+\beta)\varepsilon_3^p}(\beta+1) \end{cases} \tag{12}$$

where

$$\sigma_1 \varepsilon_1^p + \sigma_2 \varepsilon_2^p + \sigma_3 \varepsilon_3^p = C \tag{13}$$

$$\beta = \frac{\sigma_2 - \sigma_3}{\sigma_1 - \sigma_3}, \quad \left(0 \leq \beta \leq 1\right) \tag{14}$$

$$\sigma_0 = \frac{1}{3}\left(\sigma_1 + \sigma_2 + \sigma_3\right) \tag{15}$$

$$\varepsilon_0^p = \frac{1}{3}\left(\varepsilon_1^p + \varepsilon_2^p + \varepsilon_3^p\right) \tag{16}$$

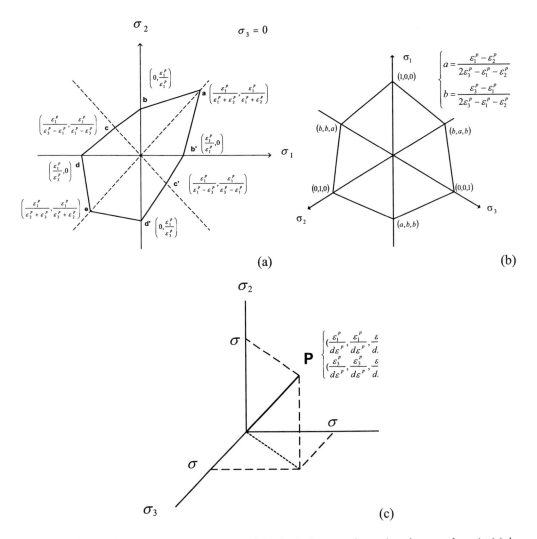

Figure 2 Yield surface of a SMA in some special principal stress planes (maximum scheme). (a) in $(\sigma_1 - \sigma_2, \ \sigma_3 = 0)$ -plane; (b) in $(\sigma_1 + \sigma_2 + \sigma_3 = 1)$ -plane; (c) position of apex.

As an alternative approach, the yield surface in some special stress planes and the apex can be found graphically as illustrated in Figure 2. Note that, $d\varepsilon^p = \varepsilon_1^p + \varepsilon_2^p + \varepsilon_3^p$, and the apex P is at $(\dfrac{\varepsilon_1^p}{d\varepsilon^p}, \dfrac{\varepsilon_1^p}{d\varepsilon^p}, \dfrac{\varepsilon_1^p}{d\varepsilon^p})$, or $(\dfrac{\varepsilon_3^p}{d\varepsilon^p}, \dfrac{\varepsilon_3^p}{d\varepsilon^p}, \dfrac{\varepsilon_3^p}{d\varepsilon^p})$ if the yield stress in uni-axial tension does not exist. If $d\varepsilon^p = 0$, i.e., without any volume change at all in the phase transformation, the position of the apex P is at infinity, which corresponds to a cylindrical shape of yield surface.

As we can see, Eqn. (11) reveals that for a SMA with a principal phase transformation strain in a format like $\omega(1,0,-1)$, where ω is a constant, the yield surface resulted by the maximum scheme is identical to the Tresca yield surface. On the other hand, not all principal phase transformation strains in a format like $\omega(1,0,-1)$ can reproduce the von Mises yield surface. According to Huang and Gao (2004), in the $(\varepsilon_{11}^p, \varepsilon_{22}^p, \varepsilon_{33}^p)$ space, these strains that satisfy the von Miese yield surface form a continuous curve (plus a separated single point). But they are discontinuous in the $(\varepsilon_{11}^p, \varepsilon_{22}^p, \varepsilon_{33}^p, \varepsilon_{12}^p, \varepsilon_{23}^p, \varepsilon_{31}^p)$ space.

Two-Pole Phenomenon

If there is only one yield mechanism in a material, according to the classic theory of plasticity, the yield surface in the $(\sigma_1, \sigma_2, \sigma_3)$ space is a slant column (refer to Figure 3a) if this material is non-compressible, i.e., without any volume change in the plastic deformation. Otherwise, it is a slant cone as illustrated in Figure 3b). In any case, it is impossible that the yield surface has two poles simultaneousely.

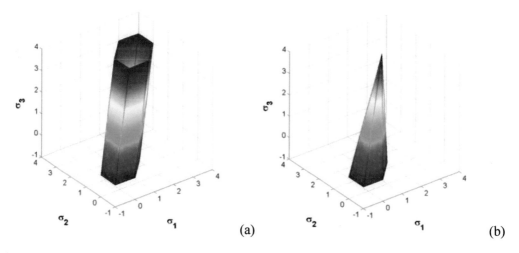

Figure 3 Yield surfaces of a (a) non-compressible material (Tresca criterion) and a (b) compressible (expandable) material (Mohr-Coulomb criterion) in the $(\sigma_1, \sigma_2, \sigma_3)$ space.

If the difference in Young's moduli of austenite and martensite of a SMA cannot be ignored, Eqn. (1) should be replaced by (Huang 2004)

$$-\tfrac{1}{2}\Sigma_{ij}(C_{ijkl}^{A}-C_{ijkl}^{M^{i}})\Sigma_{kl}+\Sigma_{ij}E_{ij}^{p}=K \tag{17}$$

where Σ and $\mathbf{E}^{p}$ are the stress tensor and phase transformation strain tensor, respectively. C_{ijkl}^{A} and $C_{ijkl}^{M^{i}}$ are the elastic moduli of austenite and martensite variant M^{i}, respectively. The following expressions may be taken for the elastic moduli of austenite and martensite (Gao and Huang 2002)

$$[C_{ijkl}^{A}]=\frac{1}{D_{A}}\begin{bmatrix}1 & -v & -v\\ -v & 1 & -v\\ -v & -v & 1\end{bmatrix}, \quad [C_{ijkl}^{M^{i}}]=\frac{1}{D_{M}}\begin{bmatrix}1 & -v & -v\\ -v & 1 & -v\\ -v & -v & 1\end{bmatrix} \tag{18}$$

where v is the Poisson ratio (assuming that the Poisson ratios of martensite and austenite are the same), and D_{A} and D_{M} are the Young's moduli of austenite and martensite, respectively.

For simplicity, as mentioned above, one of the well known approaches in the classic theory of plasticity is to assume that all grains are in series. As such, yielding will start from one grain, which requires the lowest yield stress. If we follow this approach, i.e., the maximum scheme, and assume that the elastic moduli of austenite and martensite are isotropic (once again for simplicity), it can be proved that the yield criterion for a non-textured polycrystalline SMA can be expressed as

$$\alpha^{2}\frac{1}{2}\left(\frac{1}{D_{M}}-\frac{1}{D_{A}}\right)\left[\sigma_{1}^{2}+\sigma_{2}^{2}+\sigma_{3}^{2}-2v\left(\sigma_{1}\sigma_{2}+\sigma_{2}\sigma_{3}+\sigma_{3}\sigma_{1}\right)\right]$$
$$+\alpha\left(\sigma_{1}\varepsilon_{1}^{p}+\sigma_{2}\varepsilon_{2}^{p}+\sigma_{3}\varepsilon_{3}^{p}\right)=K \tag{19}$$

Given a particular stress state $\left(\sigma_{1},\sigma_{2},\sigma_{3}\right)$, the exact yield stress can be defined as $\alpha\left(\sigma_{1},\sigma_{2},\sigma_{3}\right)$, where α is a constant. Since austenite SMAs are normally harder than that of martensite, $D_{A}>D_{M}$.

According to Eqn. (19), the pole of the yield surface of a SMA (where the stress state is $\sigma_{1}=\sigma_{2}=\sigma_{3}$) can be found by solving the following equation,

$$\frac{3}{2}\alpha^{2}\sigma_{p}^{2}(1-2v)\left(\frac{1}{D_{M}}-\frac{1}{D_{A}}\right)+\alpha\sigma_{p}d\varepsilon^{p}=K \tag{20}$$

Here, $\alpha\sigma_{p}(1,1,1)$ is the location of the pole. Since Eqn. (20) is a second order polynomial function, there should be two solutions, i.e., there are two poles in the yield surface. It is obvious that if no volume change involved in a phase transformation, i.e., $d\varepsilon^P = 0$, the two poles are symmetrical about the origin.

Figure 4 is a schematic diagram of a yield surface in the $(\sigma_1 = \sigma_2, \sigma_3)$ plane. It shows that if the variation in Young's modulus is ignored, there is no pole in the yield surface if $d\varepsilon^P = 0$ (non-compressible material). Otherwise, we can find one pole at the point a or a'. The exact location of the pole can be determined by solving Eqn. (20), but without the first term on the left side. However, if the Young's modulus is not a constant, there will be two poles located at f and g, which are two solutions of Eqn. (20).

Using the phase transformation strain of CuZnAl reported in (Lexcellent et al. 2002), the yield surface of non-textured CuZnAl polycrystal can be produced.

Figure 5 is resulted by assuming $D_A = 4D_M$ and $v = 0.33$. The calculation follows the maximum scheme. As expected, the yield surface looks like a needle with two poles.

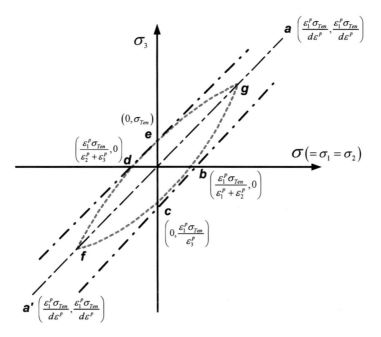

Figure 4 Schematic diagram of yield surface in the $(\sigma_1 = \sigma_2, \sigma_3)$ plane. σ_{Ten} is the yield start stress of uni-axial tension.

Although the above discussion is based on a few simplifications, the conclusion of two poles in the yield surface of SMAs should be generally valid. However, as we can see in Figure 5, very high *hydrostatic stresses* in both tri-axial tension and tri-axial compression are required in order to produce these two-poles. Practically speaking, it might be difficult for experimental verification.

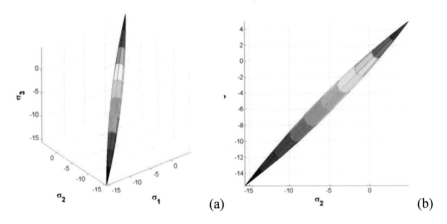

Figure 5 Yield surface of non-textured polycrystalline CuZnAl SMA. Assume that the yield stress of uni-axial tension is 1, $D_A = 4D_M$, and $v = 0.33$. (a) In the $(\sigma_1, \sigma_2, \sigma_3)$ space; (b) projection in the (σ_2, σ_3) plane.

Multiphase Transformation

One unusual feature in some SMAs is that depending on the stress state, different phase transformation may occur. That is to say, different martensite structures (not variants, which are crystallographically equivalent, but have different orientation and different plane of shearing) may be induced from the same austenite. A well-known case is β_1 CuAlNi, which may transform into either γ_1' or β_1' depending on the applied stress state. Figure 6(a) shows that their corresponding yield surfaces are apparently different. It is possible that at a temperature, the required driving energy for each transformation is not the same. Hence, the real yield surface may be obtained by re-scaling each surface according to their exact driving energies, and then applying superposition method. A schematic plot is shown in Figure 6(b). According to Figure 6(b), upon tension, the transformation is $\beta_1 \to \beta_1'$; while upon compression, it is $\beta_1 \to \gamma_1'$. This is exactly the same phenomenon observed many years ago by Otsuka and Shimizu (1982).

Above discussion indicates a possibility, at least in theory, while verification is still necessary. Recently, the yield surfaces of two non-textured polycrystalline SMAs, namely, CuAlBe and CuZnAl in the $(\sigma_1 - \sigma_2, \sigma_3 = 0)$ plane were reported in Lexcellent et al (2002). Their corresponding transformation stretches of the different possible unit cells are listed in Table 1 and Table 2, respectively. As we can see in Tables 1 and 2, theoretically there are three types of possible transformations, namely, M18R, 6M1 and 6M2, in both of them. Their corresponding phase transformation strains can be obtained following some well-established approaches, for instance, the one proposed by Lexcellent et al (2002).

At this point, we assume that for a particular SMA, the driving energies of all types of phase transformations are the same. Both average and maximum schemes are applied in the course of this study.

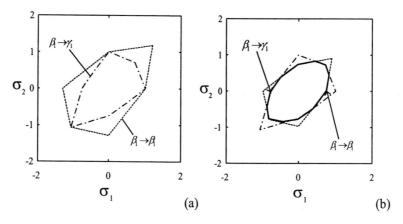

Figure 6 Yield surface of CuAlNi shape memory alloy. Solid line: combined surface; dash-dotted line: $\beta_1 \to \gamma_1'$; dotted line: $\beta_1 \to \beta_1'$. (Huang and Zhu 2002)

Figure 7 shows the resultant yield surfaces of CuAlBe and CnZnAl against the experimental results reported by Lexcellent et al (2002). The results of the maximum scheme are in black, while these of the average scheme are in grey. It reveals that the experimental results of both SMAs in the biaxial tensile stress regime are very close to the analytical results of all three types of transformations, while in the biaxial compression stress regime, 6M1 and M18R are seemingly to be the dominant types.

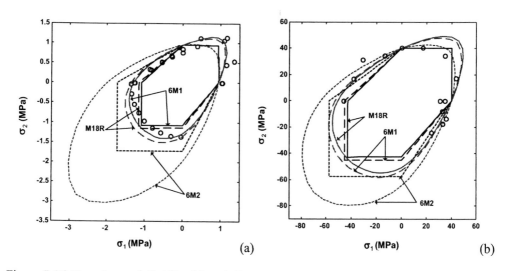

Figure 7 Yield surfaces of CuAlBe (a) and CuZnAl (b). Black: maximum scheme; grey: average scheme.

Figure 8 plots the volume increase in percentage in individual transformation of both SMAs. It is apparent that the volume increase in 6M2 of both alloys is much higher than the other types of transformation. As such, 6M2 is very like to occur upon tension, while in compression, it is the least.

In Lexcellent (2002), 6M1 was chosen for CuZlAl, while M18R was selected for CuAlBe. The selection of 6M1 for CuZnAl was based on the analysis that the 6M1 unit cell most closely describes the compatibility between austenite and a single variant of martensite

(Hane 1999). However, in a very recently published paper by Lexcellent and Blanc (2004), it was found that the stress free condition might not be a pre-requirement in the austenite-martensite inter-phase in NiTi SMAs. As the stress free condition is a well-accepted hypothesis for decades, the significance of this finding, if it can be verified by cross-examination, will be tremendous. As such, in SMAs, for instance, CuAlBe and CuZnAl, where the exact transformation upon mechanical loading depends on the applied stress state, their yield surfaces have to be resulted by the superposition method. That is to say, one cannot take it for granted that the 6M2 transformation should be ignored.

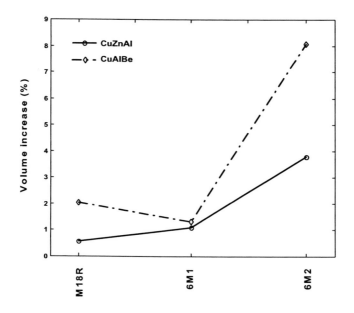

Figure 8 Variation in volume upon phase transformation.

Table 1 Transformation stretches of different possible unit cells of Cu-23.73Zn- 9.4Al (at%).

Common parameters	Unit cells		
	M18R	6M1	6M2
$\alpha = 1.070$ $\beta = 0.908$	$\gamma = 1.0$	$\gamma_1 = 1.0$	$\gamma_2 = 1.0$

Table 2 Transformation stretches of different possible unit cells of Cu-11.6Al- 0.5Be (wt%).

Common parameters	Unit cells		
	M18R	6M1	6M2
$\alpha = 1.083$ $\beta = 0896$	$\gamma = 1.0$	$\gamma_1 = 1.0$	$\gamma_2 = 1.0$

Yield Surface in Detwinning

Now we turn to another type of transformation in SMAs, namely, martensite reorientation. For simplicity, we only consider the simplest case, i.e., detwinning upon stressing in an initially pure twinned martensite. To the best knowledge of the authors, this has never been addressed in the literature before.

In this case, one may still follow the same schemes as described above. However, since no volume change is involved in the martensite reorientation, the associated transformation strain in reorientation (ε_l^r) may be written as

$$\varepsilon_l^r = \varepsilon_l^p - \frac{d\varepsilon^p}{3}\mathbf{I} \tag{21}$$

where $\mathbf{I}$ is a unit matrix.

Following the framework presented above, the yield surfaces in the (σ_1-σ_2, σ_3=0) plane of two non-textured SMAs, namely CuAlBe and CuZnAl, are determined (Figure 9). The transformation strains used in the calculation are again from Lexcellent et al (2002). For a clearer view, only one type of transformation is presented here. For CuAlBe, that is M18R, and that for CuZnAl is 6M1.

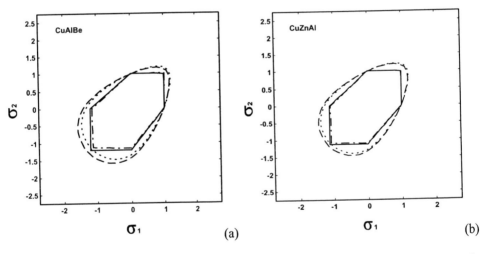

Figure 9 Yield surfaces of CuAlBe and CuZnAl in the (σ_1-σ_2, σ_3=0) plane. Solid line: phase transformation (maximum scheme); dash-dotted line: re-orientation (maximum); dashed line: phase transformation (average scheme); dotted line: reorientation (average scheme).

Figure 9 reveals that the difference in yield surface in the (σ_1-σ_2, σ_3=0) plane between the phase transformation and martensite reorientation is very small. However, as volume change is normally involved in the phase transformation in SMAs, according to Figure 2, the yield surface of the phase transformation should be a slant cone, while that of detwinning should be a slant column. Referring to Figure 8, apparently CuZnAl and CuAlBe have different shapes of yield surface for the phase transformation and de-twinning (martensite reorientation).

Conclusions

In this paper, three important issues about the yield surface of SMAs are investigated.

- The first is the two-pole phenomenon. We show that this phenomenon can happen in a SMA if there is a variation in the Young's Modulus upon the phase transformation.
- The second is about multiphase transformation in some SMAs. In such a case, the yield surface may be determined by the superposition method. We compare our predictions with that of the experimental results of two SMAs, namely, CuZnAl and CuAlBe, reported in the literature.
- The last one is the difference in the yield surface between the phase transformation and martensite reorientation. It reveals that despite that there is not much difference in the yield surface between the phase transformation and martensite reorientation in the (σ_1-σ_2, σ_3=0) plane, the yield surface of the phase transformation is a slant cone if there is a volume variation in the phase transformation, while that of the martensite reorientation is always a slant column as there is no volume change in the martensite reorientation.

References

[1] Funakubo, H. (1987). Shape memory alloys. Gordon and Breach Science Publishers, London.

[2] Gao, X.Y. and Huang, W.M. (2002). Transformation start stress in non-textured shape memory alloys. *Smart Materials and Structures*. Vol. 11, pp256-268.

[3] Hane, K.F. (2002), Bulk and thin film microstructures in untwined martensite, *Journal of the Mechanics and Physics of Solids*, Vol. 47, pp1917-1939.

[4] Huang, W. (1999a). "Yield" surfaces of shape memory alloys and their applications. *Acta Materialia*. Vol. 47, pp2769-2776.

[5] Huang, W. (1999b). Modified Shape Memory Alloy (SMA) Model for SMA Wire-Based Actuator Design. *Journal of Intelligent Material Systems and Structures*. Vol. 10, pp221-231.

[6] Huang, W.M. (2002). On the selection of shape memory alloys for actuators. *Materials and Design*. Vol. 23, pp11-19.

[7] Huang, W.M. (2004). On the effects of different Young's moduli in phase transformation start stress vs. temperature relationship of shape memory alloys. *Scripta Materialia*. Vol. 50, pp353-357.

[8] Huang, W.M. and Gao, X.Y. (2004). Tresca and von Mises yield criteria: a view from strain space. *Philosophical Magazine Letters*. Vol. 84, pp625-630.

[9] Huang, W.M. and Zhang, W.H. (2003). Surface relief phenomenon in a NiTi shape memory alloy rod. *Proceedings of SPIE*. Vol. 5277, pp198-205.

[10] Huang, W.M. and Zhu, J.J. (2002). To predict the behavior of shape memory alloys under proportional load. *Mechanics of Materials*. Vol. 34, pp547-561.

[11] Lexcellent C., Vivet A., Bouvet C., Calloch S., and Blanc P. (2002). Experimental and numerical determinations of the initial surface of phase transformation under biaxial

loading in some polycrystalline shape-memory alloys. *Journal of the Mechanics and Physics of Solids*. Vol. 50, pp2717-2735.

[12] Lexcellent, C. and Blanc P. (2004). Phase transformation yield surface determination for some shape memory alloys. *Acta Materialia*, Vol. 52, No. 8, pp2317-2324.

[13] Mura, T. (1987). *Micromechanics of defects in solids*. Martinus Nijhoff Publishess, Dordrecht.

[14] Otsuka, K. and Shimizu, K. (1982). Stress-induced martensitic transformations and martensite-to-martensite transformations, In: Aaronson, H.I., Wayman, C.M. (Eds.). *Proceedings of an International Conference on Solids → Solids Phase Transformations*, Pittsburg. Pp1267-1286.

[15] Otsuka, K. and Wayman, C.M. (1998). *Shape memory materials*. Cambridge University Press, Cambridge; New York.

[16] Taylor, G.I. (1938). Plastic strain in metals. *Journal of Institute of Metals*. Vol. 62, pp307-324.

[17] Zhu, J., Liang, N., Huang, W., Liew, K.M. and Liu, Z. (2002). A thermodynamic constitutive model for stress induced phase transformation in shape memory alloys. *International Journal of Solids and Structures*. Vol. 39, pp741-763.

In: Trends in Materials Science Research
Editor: B.M. Caruta, pp. 117-146

ISBN: 1-59454-367-4
© 2006 Nova Science Publishers, Inc.

Chapter 5

RECENT ADVANCES IN PHOTOLUMINESCENCE OF AMORPHOUS CONDENSED MATTER

Jai Singh

School of Engineering and Logistics, Faculty of Technology, B-41, Charles Darwin University, Darwin, NT 0909, Australia

Abstract

A comprehensive review of the recent developments in studying the photoluminescence in amorphous materials is presented. Experimental results are reviewed and new theoretical results are derived. Four possibilities of radiative recombination for an exciton are considered: (i) both of the excited electron and hole are in their extended states, (ii) electron is in the extended and hole in tail states, (iii) electron is in the tail and hole in extended states and (iv) both in their tail states. Rates of radiative recombination corresponding to each of the four possibilities are derived: a) within two-level approximation, and at (b) non-equilibrium and (c) equilibrium conditions. It is found that the rates derived under the non-equilibrium condition, have no finite peak values with respect to the photoluminescence energy. Considering that the maximum value of the rates derived at equilibrium gives the inverse of the radiative lifetime, the latter is calculated for all the four possibilities in a-Si:H. The theory is general and is expected to be applicable to all amorphous materials, including organic materials used for fabricating light emitting devices.

Introduction

The photoluminescence (PL) in hydrogenated amorphous silicon (a-Si:H) has been studied extensively in the last two decades [1-5] because it provides direct information about the electronic states and carrier dynamics in the material. Both, evolution of PL peak structure with the decay time and time resolved behaviour, have been studied [6-7]. Using the time-resolved spectroscopy (TRS), Wilson et al. [6] have observed PL peaks with the radiative lifetime in the nanosecond (ns), microsecond (μs) and millisecond (ms) time ranges in a-Si:H at a temperature of 15 K. In contrast to this, using the quadrature frequency resolved spectroscopy (QFRS), other groups [8-11] have observed only a double peak structure PL in a-Si:H at the liquid helium temperature. One peak appears at a short time in the μs range and the other in the ms range [7,12]. Results of a recent measurement [13] of the normalized QFRS PL signals versus PL lifetime in a-Si:H at different PL energies are shown in Fig. 1, which clearly illustrates the appearance of two peaks. Thus, recent QFRS measurments do not exhibit any PL peak in the nanosecond region in a-Si:H. The QFRS measurements can be carried out at low excitation densities, what is also called the geminate recombination limit, and provide information on the lifetime distribution without deconvolution. Therefore, according to the QFRS measurements, the faster peak has been attributed to the radiative recombination of singlet excitons and slower peak to that of triplet excitons [7,12]. This is based on the fact that the recombination of singlet excitons is spin-allowed and hence faster than the recombination of triplet excitons, which is spin-forbidden. A similar double peak structure with one fast peak in the μs region and slow peak in the ms region has also been observed in hydrogenated amorphous germanium (a-Ge:H). Using the effective mass approach, a theory for the excitonic states in amorphous semiconductors has been developed by Singh et al. [5] and the occurrence of the double peak structure has successfully been explained. The theory also enables one to calculate the energy difference between the singlet and triplet states and it has been successfully applied to both a-Si:H and a-Ge:H.

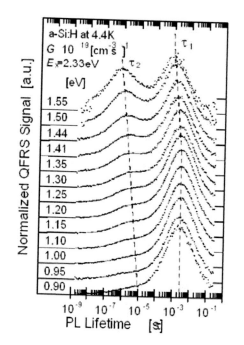

Figure 1: QFRS spectra of monochromatized photoluminescence for $\hbar\omega$ ranging from 0.95 to 1.55 eV in a-Si:H observed at 4.4 K with an excitation energy, E_{ex} = 2.33 eV and at a generation rate G = $10^{19} cm^{-3} s^{-1}$ [13].

For applying the TRS technique, one has to excite the sample with a very high excitation density. For example, the excitation density used by Wilson et al. [3] is $10^{19} cm^{-3}$ whereas in QFRS experiments Aoki et al. [7] have used much lower excitation density of 10^{11} - 10^{13} cm^{-3}. The difference in the excitation density between TRS and QFRS experiments may be used to explain the appearance of a ns peak in TRS experiment but not in QFRS experiments. At a very high excitation density, strong repulsive interactions between like charge carriers may slow down the relaxation of excitons to the tail states. In this situation, excitons may dominantly recombine radiatively from the extended states before relaxing to the tail states. From this view point, one may argue that the reason for not observing the ns peak in QFRS experiments is the low excitation density that allows excitons to relax down to the tail states before recombining radiatively. This also raises a question: Is the radiative recombination from the extended states faster and in the ns range and that from the tail states is slower and in the μs range?

Recently Aoki's group [13-15] has used as high excitation densities in their QFRS experiments as used by Wilson et al. [3] but still they have not observed any peak in the ns range in a-Si:H at 4.4 K. This suggests that the excitation density may not be the reason for not observing any ns peak in QFRS experiments. Another problem is that although there are several studies on the radiative recombination in crystalline solids [16-18], there are very few theoretical studies [19] in amorphous semiconductors. Moreover, these earlier works [19] have not considered excitonic radiative recombination and therefore one cannot answer the questions raised above, namely: 1) is the excitonic radiative recombination from the extended states three orders of magnitude faster than that from the tail states and 2) does the radiative lifetime depend on the excitation density?

In this chapter, the recent theoretical developments in calculating the rate of spontaneous emission and radiative lifetime of excitons in amorphous semiconductors (a-semiconductors) are presented. A fundamental assumption in these developments is that an exciton can be formed between an excited pair of electron (e) and hole (h) by a photon of energy equal to or higher than the optical gap energy so that initially both the charge carriers are excited in their extended states. As an exciton so created relaxes downward to tails states, it remains an exciton with its identified excitonic Bohr radius and binding energy until the charge carriers recombine radiatively by emitting a photon. Thus, the excitonic relaxation is restricted by the excitonic internal energy quantum states and therefore it is not as fast as the thermal relaxation of free carriers (not bound in excitons). This results in a peaked excitonic photoluminescence (see, e.g., Fig. 1) as has been observed in several experiments [3,4,6-15]. Four possibilities are considered: (i) both of the excited electron and hole are in their extended states, (ii) electron is in the extended and hole in tail states, (iii) electron is in the tail and hole in extended states and (iv) both in their tail states. We have first derived the rates of spontaneous emission within two-level approximation, and then for the four possibilities we have derived them under both non-equilibrium and equilibrium conditions. It is found that the rates derived within the two-level approximation and under non-equilibrium are not applicable for studying the PL radiative lifetime in amorphous semiconductors. Rates derived under equilibrium are used to calculate the PL radiative lifetime in a-Si:H associated with all the four possibilities. It is shown that the rates depend on the initial energy state of recombination, emitted photon energy and temperature. Rates are found to be one to two orders of magnitude faster in the possibilities (i) – (iii) than those in the possibility (iv). It is known that at low temperatures, both singlet and triplet states can be formed and the shorter lifetime in the μs range is due to the radiative recombination of singlet excitons and the longer lifetime in the ms range is due to that of triplet excitons.

In sections 2-3, rates of spontaneous emission are first derived within the two-level-approximation, and then under non-thermal equilibrium and thermal equilibrium for the above mentioned four possibilities of radiative recombinations. The theory developed under equilibrium is then applied to a-Si:H and results thus obtained are presented in section 4. A discussion of the theory and results is given in section 5.

Radiative Recombination

We consider an exciton excited such that its electron (e) is in the conduction states and hole (h) in the valence states, and then they recombine radiatively by emitting a photon due to the exciton-photon interaction. The interaction operator between a pair of excited e and h and a photon can be written as:

$$\hat{H}_{xp} = -(\frac{e}{m_e^*}\mathbf{p}_e - \frac{e}{m_h^*}\mathbf{p}_h) \cdot \mathbf{A},$$ (1)

where m_e^* and $\mathbf{p}_e$ and m_h^* and $\mathbf{p}_h$ are the effective masses and linear momenta of the excited electron and hole, respectively, and $\mathbf{A}$ is the vector potential given by:

$$\mathbf{A} = \sum_\lambda \left(\frac{\hbar}{2\varepsilon_0 n^2 V \omega_\lambda}\right)^{1/2} \left[c_\lambda^+ \hat{\varepsilon}_\lambda + c.c.\right],$$ (2)

where n is the refractive index and V volume of the material, ω_λ is the frequency and c_λ^+ is the creation operator of a photon in a mode λ, and $\hat{\varepsilon}_\lambda$ is the unit polarization vector of photons.. The second term of $\mathbf{A}$ corresponds to the absorption and will not be considered here onward.

Using the centre of mass , $\mathbf{R}_x = \frac{m_e^* \mathbf{r}_e + m_h^* \mathbf{r}_h}{M}$ and relative $\mathbf{r} = \mathbf{r}_e - \mathbf{r}_h$ coordinate transformations, the interaction operator $\hat{H}_{xp}$ [Eq. (1)] can be transformed into:

$$\hat{H}_{xp} = -\frac{e}{\mu_x}\mathbf{A} \cdot \mathbf{p},$$ (3)

where $\mathbf{p} = -i\hbar \nabla_r$ is the linear momentum associated with the relative motion between e and h in an exciton and μ_x is the excitonic reduced mass ($\mu_x^{-1} = m_e^{*-1} + m_h^{*-1}$). The operator in Eq. (3) does not depend on the centre of mass motion of the exciton. It may be noted here that the operator in Eq. (1) is the same for the exciton-photon interaction or a pair of e and h and photon interaction. This is because the recombination between a pair of e and h occurs due to their Coulombic attraction, which always exists whether they form an exciton or not. This is the reason that the operator is independent of the centre of mass momentum.

For amorphous solids, it is important to distinguish whether the excited charge carriers are created in the extended states or tail states [5,20-23]. There are four possibilities: i) both excited e and h are in their respective extended states, (ii) e in its extended and h in its tail states, (iii) e in its tail and h in its extended states and (iv) both e and h are in their respective tail states. For studying PL, it is important to identify which one of the four possibilities we are dealing with. This is because charge carriers have different wave functions, effective masses, and hence different excitonic Bohr radii in their extended and tail states [5,20-23].

It may be emphasized here that the thermal relaxation of the excited free charge carriers from extended to tail states occurs in the picosecond (ps) time scale. That means the majority of excited carriers should relax down to the tail states before recombining radiatively and giving rise to PL, which occurs in the ns time scale or longer. In this case, it is only meaning full to consider the possibility (iv) for studying PL in a-semiconductors. This is true, however, only for the radiative recombination of the free excited charge carriers, which do not form excitons. This is because in an exciton, the excited electron and hole are bound through their Coulomb interaction in hydrogen like energy states of a certain excitonic Bohr radius and energy, and hence cannot thermalise like free excited charge carriers. They can only relax down non-radiatively by maintaining their Bohr radius and internal energy and radiatively through quantum transitions. Therefore for the excitonic radiative recombination, all four possibilities are relevant to consider for the general development of the theory. This is also supported by the observed Stokes shifts in a-Si:H [3-4,13-14], which will be discussed here later on.

The field operator $\hat{\psi}_c(\mathbf{r}_e)$ of an electron in the conduction states can be written as:

$$\hat{\psi}_c(\mathbf{r}_e) = N^{-1/2} \sum_l \exp(it_e \cdot \mathbf{R}_l^e) \phi_l(\mathbf{r}_e) a_{cl} \, , \tag{4}$$

where N is the number of atoms in the sample, $\mathbf{R}_l^e$ is the position vector of an atomic site at which the electron is created, $\phi_l(\mathbf{r}_e)$ is the wave function of an electron at the excited site l, $\mathbf{r}_e$ is the position coordinate of the electron with respect to site l and $\mathbf{t}_e$ is given by:

$$|\mathbf{t}_e| = t_e = \sqrt{2m_e^*(E_e - E_c)}/\hbar \, , \tag{5}$$

where E_e is the energy of the electron and E_c is that of its mobility edge. a_{cl} is the annihilation operator of an electron with energy E_e at a site l in the conduction c states. According to Eq. (5), if the electron energy E_e is above the mobility edge, then the electron moves as a free particle in the conduction extended states, but if $E_e < E_c$ then the electron gets localized because t_e becomes imaginary and the envelope function becomes exponentially decreasing.

Likewise the field operator, $\hat{\psi}_v(\mathbf{r}_h)$, of a hole excited with an energy E_h in the valence states can be written as:

$$\hat{\psi}_v(\mathbf{r}_h) = N^{-1/2} \sum_l \exp(-it_h \cdot \mathbf{R}_l^e) \phi_l(\mathbf{r}_h) d_{vl} \, , \quad d_{vl} = a_{vl}^+ \, , \tag{6}$$

where

$$|\mathbf{t}_h| = t_h = \sqrt{2m_h^*(E_v - E_h)}/\hbar \,, \qquad (7)$$

where E_v is the energy of the hole mobility edge, and d_{vl} is the annihilation operator of a hole in the valence states, v, with energy E_h. Here again the hole behaves like a free particle for $E_v > E_h$ but gets localized in the tail states for $E_v < E_h$.

Using Eqs. (2), (4) and (6), the interaction operator $\hat{H}_{xp}$ [Eq. (3)] can be written in the second quantized form as:

$$\hat{H}_{xp} = -\frac{e}{\mu_x} \sum_{\lambda} \left(\frac{\hbar}{2\varepsilon_0 n^2 V \omega_\lambda} \right)^{1/2} Q_{cv} c_\lambda^+ \,, \qquad (8)$$

where

$$Q_{cv} = N^{-1} \sum_l \sum_m \exp[-i\mathbf{t}_e \cdot \mathbf{R}_l^e] \exp[-i\mathbf{t}_h \cdot \mathbf{R}_m^h] Z_{lm\lambda} B_{cvlm} \,, \qquad (9)$$

where $B_{cvlm} = a_{cl} d_{vm}$ is the annihilation operator of an exciton by annihilating an electron in the conduction c states at site l and a hole in the valence v states at site m and $Z_{lm\lambda}$ is given by:

$$Z_{lm\lambda} = \int \phi_l^*(r_e) \hat{\varepsilon}_\lambda \cdot \mathbf{p} \phi_m(r_h) dr_e dr_h \,. \qquad (10)$$

It may be noted that it is not necessary to use the exciton operator B_{cvlm} in Eq. (9), one can also use the product of fermion operators $a_{cl} d_{vm}$ instead.

We now consider a transition from an initial state with one exciton created at a site, say, l (both e and h created on the same site) any other site is assumed to have zero excitons. We also assume that there are no photons in the initial state. The initial state, $|i>$ can be expressed in terms of the occupation number as: $|i> = |0, 0, \dots 1_{(cl,\, vl)}, 0, 0, \dots,0>|0> = |1>|0>$. The final state is assumed to have no excitons, but only one photon in a mode λ and can be expressed as: $|f> = |0>|0, 0, \dots, 1_\lambda, 0, 0, \dots,0> = |0>|1>$. The transition matrix element is then obtained as:

$$<f|\hat{H}_{xp}|i> = -\frac{e}{\mu_x} \sum_{\lambda} \left(\frac{\hbar}{2\varepsilon_0 n^2 V \omega_\lambda} \right)^{1/2} p_{cv} \,, \qquad (11)$$

where

$$p_{cv} = N^{-1} \sum_l \sum_m \exp[-it_e \cdot \mathbf{R}_l^e] \exp[-it_h \cdot \mathbf{R}_m^h] Z_{lm\lambda} \delta_{l,m} \ . \tag{12}$$

The derivation of p_{cv} depends on where the radiative recombination originates from e.g., extended or tail states. Therefore, it is important to consider the four possibilities, i) both e and h are in their extended states, ii) e in the extended and hole in the tail states, iii) e in the tail and hole in the extended states and iv) both e and h are in their tails states, because the envelope functions of electron and hole are different in the extended and tail states.

Let us first derive $Z_{lm\lambda}$. There are two approaches, which are used in amorphous solids to evaluate this integral. The integral actually determines the average value of the relative momentum between the excited electron-hole pair. In the first approach, it is assumed to be a constant and independent of the photon energy as [22,24]:

$$Z_{lm\lambda} = \int \phi_l^*(r_e) \hat{\varepsilon}_\lambda \cdot \mathbf{p} \phi_m(r_h) d\mathbf{r}_e d\mathbf{r}_h = Z_1 = \pi \hbar \left(\frac{L}{V} \right)^{1/2} , \tag{13}$$

where L is the average bond length in a sample and Z_1 denotes the matrix element obtained from the first approach.

In the second approach, the integral is evaluated using the dipole approximation as [20-23, 25]:

$$Z_{lm\lambda} = \int \phi_l^*(r_e) \hat{\varepsilon}_\lambda \cdot \mathbf{p} \phi_m(r_h) d\mathbf{r}_e d\mathbf{r}_h = Z_2 = i\omega \mu_x r_{eh} , \tag{14}$$

where Z_2 denotes the transition matrix element obtained from approach 2, $\hbar\omega = E_c' - E_v'$ is the emitted photon energy and $r_{eh} = <l \mid \hat{\varepsilon}_\lambda . \mathbf{r} \mid m>$ is the average separation between the excited electron-hole pair, which also can be assumed to be site independent. Thus, in both the approaches the integral becomes site independent and can be taken out of the summation.

In the case of excitons, it can be easily assumed that $r_{eh} = a_{ex}$, the excitonic Bohr radius. As far as the transition matrix element is concerned, this is the only difference between an excitonic recombination and free e and h or distant pair recombination. In case of the former $r_{eh} = a_{ex}$, the excitonic Bohr radius but in case of the latter r_{eh} is the average separation between e and h.

It may be noted here that the value of the integral obtained from the first approach, Z_1, leads to the well known Tauc's relation for the absorption coefficient observed in amorphous semiconductors and that obtained from the second approach is used to explain the deviations from Tauc's relation in the absorption coefficient observed in some amorphous semiconductors [20-23,25].

Using Eqs. (13) and (14), p_{cv} in Eq. (12) can be written as:

$$p_{cv} = Z_i N^{-1} \sum_l \sum_m \exp[-it_e \cdot \mathbf{R}_l^e] \exp[-it_h \cdot \mathbf{R}_m^h] \delta_{l,m}' , i = 1, 2 \ . \tag{15}$$

Now the derivation of p_{cv} in Eq. (15) depends on the four possibilities, which will be considered below separately.

(i)Extended-to-extended states transitions
Rearranging the exponents in Eq. (15) as:

$$\exp[-it_e \cdot \mathbf{R}_l^e - it_h \cdot \mathbf{R}_m^h] = \exp[-it_e \cdot (\mathbf{R}_l^e - \mathbf{R}_m^h) - i(t_e + t_h) \cdot \mathbf{R}_m^h], \qquad (16)$$

and identifying the fact that $\mathbf{R}_l^e - \mathbf{R}_l^h = \mathbf{a}_{ex}$, the excitonic Bohr radius (the separation between e and h in an exciton prior to their recombination, which is assumed to be site independent), the first exponential becomes site independent. It can be taken out of the summation signs and then the matrix element in Eq. (15) becomes:

$$p_{cv} = Z_i \exp[-it_e \cdot \mathbf{a}_{ex}]\delta_{t_e,-t_h} , i = 1,2 , \qquad (17)$$

where $\delta_{t_e,-t_h}$ represents the momentum conservation in the transition. The square of the transition matrix element then becomes:

$$| p_{cv} |^2 = | Z_i^* Z_i | = | Z_i |^2 , i = 1,2 . \qquad (18)$$

(ii) Transitions from extended to tail states
Here we consider e excited in the extended and h in the tail states. For this case Eq. (15) can be written as:

$$p_{cv} = Z_i N^{-1} \sum_l \sum_m \exp[-it_e \cdot \mathbf{R}_l^e] \exp[-t_h' \cdot \mathbf{R}_m^h]\delta_{l,m} , \qquad (19)$$

where

$$| \mathbf{t}_h' |= t_h' = \sqrt{2m_h^*(E_h - E_v)}/\hbar . \qquad (20)$$

Rewriting Eq. (19) as:

$$p_{cv} = Z_i N^{-1} \sum_l \sum_m \exp[-it_e \cdot (\mathbf{R}_l^e - \mathbf{R}_m^h)]\exp[-i(t_e - it_h') \cdot \mathbf{R}_m^h]\delta_{l,m} , \qquad (21)$$

and simplifying it in the same way as Eq. (16), we get:

$$p_{cv} = Z_i \exp[-it_e \cdot \mathbf{a}_{ex}]\delta_{t_e,it_h'} , i = 1,2 . \qquad (22)$$

Eq. (22) gives the same expression for $|p_{cv}|^2$ as in Eq. (18) for the possibility (i), and in this case also the momenta remain conserved.

(iii) Transitions from tail-to-extended states

Here we consider e excited in the tail and h in the extended states. For this case the transition matrix element can be derived in a way analogous to that for the extended to tail states. p_{cv} is obtained as:

$$p_{cv} = Z_i \exp[-i t_h \cdot \mathbf{a}_{ex}] \delta_{it'_e, t_h}, \, i = 1,2, \tag{23}$$

where

$$| \mathbf{t}'_e | = t'_e = \sqrt{2 m^*_e (E_c - E_e)} / \hbar. \tag{24}$$

Here again $|p_{cv}|^2$ remains the same as in Eq. (18).

(iv) Transitions from tail-to-tail states

Here we consider that both e and h are excited in the tail states. For this case, using the localized form of the electron and hole wave functions, p_{cv} is obtained as:

$$p_{cv} = Z_i N^{-1} \sum_l \sum_m \exp[-\mathbf{t}'_e \cdot \mathbf{R}^e_l] \exp[-\mathbf{t}'_h \cdot \mathbf{R}^h_m] \delta_{l,m}. \tag{25}$$

Here again, one can rearrange the exponential as:

$$p_{cv} = Z_i N^{-1} \sum_l \sum_m \exp[-\mathbf{t}'_e \cdot (\mathbf{R}^e_l - \mathbf{R}^h_m)] \exp[-(\mathbf{t}'_h + \mathbf{t}'_e) \cdot \mathbf{R}^h_m], \tag{26}$$

which becomes:

$$p_{cv} = Z_i \exp[-\mathbf{t}'_e \cdot \mathbf{a}_{ex}] \delta_{t'_e, -t'_h}, \, i = 1,2. \tag{27}$$

Assuming that the relative momentum of electron, $\hbar t'_e$, will be along the direction of $\mathbf{a}_{ex}$ at the time of recombination, $|p_{cv}|^2$ from Eq. (27) becomes:

$$| p_{cv} |^2 = | Z_i |^2 \exp[-2 t'_e a_{ex}] \delta_{t'_e, -t'_h}, \, i = 1,2. \tag{28}$$

Rates of Spontaneous Emission

Using Eq. (11) and applying Fermi's golden rule, the rate R_{sp} (s^{-1}) of spontaneous emission can be written as [17-18, 26-27]:

$$R_{sp} = \frac{2\pi e^2}{\mu_x^2} \left(\frac{1}{2\varepsilon_0 n^2 V \omega_\lambda} \right) \sum_{E'_c, E'_v} | p_{cv} |^2 f_c f_v \delta(E'_c - E'_v - \hbar\omega_\lambda), \tag{29}$$

where f_c and f_v are the probabilities of occupation of an electron in the conduction and a hole in the valence states, respectively. The rate of spontaneous emission in Eq. (29) can be evaluated under several conditions, which will be described below:

Rate of Emission under Two-Level Approximation

Here only two energy levels are considered as in atomic systems. An electron takes downward transition from an excited state to the ground state; no energy bands are involved. In this case, $f_c = f_v = 1$ and then denoting the corresponding rate of spontaneous emission by R_{sp12}, we get:

$$R_{sp12} = \frac{\pi e^2}{\varepsilon_0 n^2 \omega \mu_x^2} |p_{12}|^2 \, \delta(\hbar\omega - \hbar\omega_\lambda), \tag{30}$$

where $\hbar\omega = E_2 - E_1$, which is the energy difference between the excited (E_2) and ground (E_1) states. In this case p_{12} is derived using the dipole approximation [Eq. (14)] as:

$$p_{12} = i\omega\mu_x <\hat{\varepsilon}_\lambda \cdot \mathbf{r}> , \tag{31}$$

where $\mathbf{r}$ is the dipole length, and $< ...>$ denotes integration over all photon modes λ.

Substituting Eq. (31) in Eq. (30), and then integrating over the photon wave vector $\mathbf{k}$ using $\omega = kc/n$, we get [3,26]:

$$R_{sp12} = \frac{4\kappa e^2 \sqrt{\varepsilon} \omega^3 |r|^2}{3\hbar c^3}, \tag{32}$$

where $\varepsilon = n^2$ is the static dielectric constant, $|r|$ is the mean separation between the electron and hole and $\kappa = 1/(4\pi\varepsilon_0)$.

The expression of the rate of spontaneous emission obtained in Eq. (32) is well known [3,6,26] and it is independent of the electron and hole masses and temperature. As only two discrete energy levels are considered, the density of states is not used in the derivation. Therefore, although using $|r| = a_{ex}$, the excitonic Bohr radius, the rate R_{sp12} has been applied for a-Si:H [3,6] it should only be used for calculating the radiative recombination in isolated atoms, not in condensed matter. In amorphous semiconductors, as a_{ex} is not known, Eq.(32) has been used to estimate it by equating the inverse of the rate to the radiative lifetime of PL measured experimentally [3,6-7]. This cannot be expected to give very satisfactory result.

Rates of Spontaneous Emission in Amorphous Solids

In applying Eq. (29) for any condensed matter system, it is necessary to determine f_c and f_v and the density of states. On one hand, it may be argued that the short-time photoluminescence can occur before the system reaches thermal equilibrium and therefore no equilibrium distribution functions can be used for the excited charge carriers. In this case one should use $f_c = f_v = 1$ in Eq. (29) for condensed matter as well. On the other hand, as the carriers are excited by the same energy photons, even in a short time delay, they may be expected to be in thermal equilibrium among themselves but not necessarily with the lattice. Therefore, they will relax according to an equilibrium distribution. As the electronic states of amorphous solids include the localized tail states, it is more appropriate to use the Maxwell-Boltzmann distribution for this situation. We will consider here radiative recombination under both non-equilibrium and equilibrium conditions.

A) Recombination in non-thermal equilibrium

The transition matrix elements corresponding to the four possibilities have already been derived in section 2. One gets the same expression for $|p_{cv}|^2$ for possibilities (i)-(iii) [see Eqs. (17), (22) and (23)]. Considering $f_c = f_v = 1$ and substituting $|p_{cv}|^2$ [Eq. (12)] derived above from the two approaches in Eq. (29), we get the rates, R_{spni}, for the possibilities (i) – (iii) as:

$$R_{spni} = \frac{2\pi e^2}{\mu_x^2}\left(\frac{1}{2\varepsilon_0 n^2 V \omega_\lambda}\right)|Z_i|^2 \sum_{E_c', E_v'} \delta(E_c' - E_v' - \hbar\omega_\lambda), i = 1, 2. \tag{33}$$

where the subscript spn in R_{spni} denotes rates of spontaneous emission derived at the non-equilibrium. For evaluating the summation over E_c' and E_v' in Eq. (33), the usual approach is to convert it into an integral by using the excitonic density of states, which can be obtained as follows. Using the effective mass approximation, the electron energy in the conduction band and hole energy in the valence band can be written as:

$$E_c' = E_c + \frac{p_e^2}{2m_e^*}, \tag{34}$$

and

$$E_v' = E_v - \frac{p_h^2}{2m_h^*}, \tag{35}$$

Subtracting Eq. (35) from Eq. (34) and then applying the centre of mass and relative coordinate transformations (see above Eq. (3)), we get the excitonic energy E_x in the parabolic form as:

$$E_x = E_0 + \frac{P^2}{2M} + \frac{p^2}{2\mu_x}, \tag{36}$$

where $E_0 = E_c - E_v$ is the optical gap, $\mathbf{P}$ and $M = m_e^* + m_h^*$ are linear momentum and mass of an exciton associated with its centre of mass motion and the last term is the kinetic energy of the relative motion between e and h, which contributes to the exciton binding energy through the attractive Coulomb interaction potential between them [28]. The exciton density of states then comes from the parabolic form of the second term associated with the centre of mass motion as:

$$g_x = \frac{V}{2\pi^2}\left(\frac{2M}{\hbar^2}\right)^{3/2}(E_x - E_o)^{1/2} \ . \tag{37}$$

Using Eq.(37), the summation in Eq. (33) can be converted into an integral as:

$$\sum_{E_x'}\delta(E_x' - \hbar\omega) = I_j = \int_{E_x'} g_x \delta(E_x' - \hbar\omega)dE_x' \ , \tag{38}$$

which gives:

$$I_j = \frac{V}{2\pi^2}\left(\frac{2M}{\hbar^2}\right)^{3/2}(\hbar\omega - E_0)^{1/2} \ . \tag{39}$$

Substituting Eq. (39) in Eqs. (33) we get:

$$R_{spni}^j = \frac{2\pi e^2}{\mu_x^2}\left(\frac{1}{2\varepsilon_0 n^2 V\omega_\lambda}\right)|Z_i|^2\, I_j\Theta(\hbar\omega - E_0), i = 1,2, \tag{40}$$

where $\Theta(\hbar\omega - E_0)$ is a step function used to indicate that there is no radiative recombination for $\hbar\omega < E_0$ and j denotes that these rates are derived through the joint density of states. Having derived the rates of spontaneous emission using the excitonic density of states, it is important to remember that the use of such joint density of states for amorphous semiconductors does not give the well known Tauc's relation [22, 24] in the absorption coefficient [21]. Therefore, by using it in calculating the rate of spontaneous emission, one would violate the Van Roosbroeck and Shockley relation [29] between the absorption and emission. For this reason, the product of individual electron and hole density of states is used in evaluating the summation in Eq. (33) for amorphous semiconductors. This approach has proven to be very useful in amorphous solids as it gives the correct Tauc's relation [21-22, 24,30].

Using the product of individual density of states, the summations over E_c' and E_v' in Eqs. (33) can be evaluated by converting these into integrals as:

$$I = \int_{E_c}^{E_v + \hbar\omega} \int_{E_c' - E_v}^{E_v} g_c(E_c')g_v(E_v')\delta(E_c' - E_v' - \hbar\omega)dE_c'dE_v' \ , \tag{41}$$

where $g_c(E'_c)$ and $g_v(E'_v)$ are the densities of states of the conduction and valence states, respectively, and within the effective mass approximation these can be written as:

$$g_q(E') = \frac{V}{2\pi^2}\left(\frac{2m^*}{\hbar^2}\right)^{3/2} E_q^{1/2} \;,\; q = c, v, \tag{42}$$

where m^* is the effective mass of the corresponding charge carrier and $q = c$ (conduction) and $q = v$ (valence) states. Substituting Eq. (42) in Eq. (41), the integral can be evaluated analytically to give [22]:

$$I = \frac{V\left(m_e^* m_h^*\right)^{3/2}}{4\pi^3 \hbar^6}(\hbar\omega - E_0)^2. \tag{43}$$

Using Eq. (43) in Eq. (33) and substituting the corresponding Z_i, we get the two rates in non-thermal equilibrium for the possibilities (i) – (iii) as:

$$R_{spn1} = \frac{e^2 L (m_e^* m_h^*)^{3/2}}{4\varepsilon_0 \hbar^3 n^2 \mu_x^2 (\hbar\omega)}(\hbar\omega - E_0)^2 \Theta(\hbar\omega - E_0), \tag{44}$$

and

$$R_{spn2} = \frac{(m_e^* m_h^*)^{3/2} e^2 a_{ex}^2 V}{2\pi^2 \varepsilon_0 n^2 \hbar^7}\hbar\omega(\hbar\omega - E_0)^2 \Theta(\hbar\omega - E_0). \tag{45}$$

For excitonic transitions, it is more appropriate to replace m_e^* and m_h^* in Eqs. (44) and (45) by the excitonic reduced mass μ_x. It may also be noted that the volume V is appearing in R_{spn2} [Eq. (45)]. This is inevitable through the second approach and it has been tackled earlier by Cody [25] by defining $2N_0/V = v\rho_A$, where N_0 is the number of single spin states in the valence band and thus $2 N_0$ becomes the total number of valence electrons occupying N_0 states, v is the number of coordinating valence electrons per atom and ρ_A is the atomic density per unit volume. Thus replacing V by V/N_0 in Eq. (45) one can get around this problem. We thus obtain:

$$R_{spn1} = \frac{e^2 L \mu_x}{4\varepsilon_0 \hbar^3 n^2 (\hbar\omega)}(\hbar\omega - E_0)^2 \Theta(\hbar\omega - E_0), \tag{46}$$

and

$$R_{spn2} = \frac{\mu_x^3 e^2 a_{ex}^2}{2\pi^2 \varepsilon_0 n^2 \hbar^7 v \rho_A} \hbar \omega (\hbar \omega - E)^2 \Theta(\hbar \omega - E_0),$$ (47)

It may be emphasized here that E_0, defined as the energy of the optical gap, is not always the same for amorphous solids. It depends on the lowest excitonic state within the extended states, which is not easy to determine; neither theoretically nor experimentally. As rates derived in Eqs. (46) and (47), do not have any peak value, it is not possible to determine E_0 from these rates.

Likewise, using $|p_{cv}|^2$ [Eq. (28)] in Eq. (33) for the possibility (iv), we get the rates of spontaneous emission from the two approaches for the tail-to-tail states transitions as:

$$R_{spnti} = R_{spni} \exp(-2t'_e a_{ex}) , i = 1, 2 ,$$ (48)

where the subscript *spnt* of R_{spnt} stands for the spontaneous emission at non-equilibrium from tail-to-tail states. a_{ex} is the excitonic Bohr radius in the tail states and it is given by[5,22]:

$$a_{ex} = \frac{5\mu\varepsilon}{4\mu_x} a_0 ,$$ (49)

where μ is the reduced mass of an electron in the hydrogen atom and $a_0 = 0.529$ Å is the Bohr radius. Results for the rates of spontaneous emission obtained above are valid in the non-thermal equilibrium condition as stated above. However, unless one knows the relevant effective masses of charge carriers and the value of E_0, these rates cannot be applied to determine the lifetime of photoluminescence. The effective mass of charge carriers can be determined, as described later on, but not E_0. For this reason, it is useful first to derive these rates under thermal equilibrium as well, as shown below:

B) Recombination in thermal equilibrium

Assuming that the excited charge carriers are in thermal equilibrium among themselves, the distribution functions, f_c and f_v, can be given by the Maxwell-Boltzmann distribution function as [31]:

$$f_c = \exp[-(E_e - E_{Fn})/\kappa_B T] ,$$ (50)

and

$$f_v = \exp[-(E_{Fp} - E_h)/\kappa_B T] ,$$ (51)

where E_e and E_h are the energies of an electron in the conduction and a hole in the valence states, respectively, and E_{Fn} and E_{Fp} are the corresponding Fermi energies. κ_B is the Boltzmann constant and T temperature of the excited charge carriers. The product $f_c f_v$ is then obtained as:

$$f_c f_v \approx \exp[-(\hbar\omega - E_0)/\kappa_B T], \tag{52}$$

where $E_e - E_h = \hbar\omega$ and $E_{Fn} - E_{Fp} = E_0$ are used. It may be noted that Eq. (52) is also an approximate form of the Fermi-Dirac distribution obtained for $E_e > E_{Fn}$ and $E_h < E_{Fp}$ and it has been used widely [17-18, 26] for calculating the rate of spontaneous emission in semiconductors. Substituting Eq. (52) in the rate in Eq. (29), the integral in Eq. (41) becomes:

$$I = \frac{V(m_e^* m_h^*)^{3/2}}{4\pi^3 \hbar^6} (\hbar\omega - E_0)^2 \exp[-(\hbar\omega - E_0)/\kappa_B T] \tag{53}$$

Using Eq. (53), the rates of spontaneous emission at the thermal equilibrium for the possibilities (i) – (iii) are obtained as:

$$R_{spi} = R_{spni} \exp[-(\hbar\omega - E_0)/\kappa_B T], \, i = 1, 2, \tag{54}$$

and for the possibility (iv), we get:

$$R_{spti} = R_{spnti} \exp[-(\hbar\omega - E_0)/\kappa_B T], \, i = 1, 2. \tag{55}$$

Rates in Eqs. (54) and (55) have a maximum value, which can be used to determine E_0 as described below.

Determining E_0

For determining E_0, we assume that the peak of the observed PL intensity occurs at the same energy as that of the rate of spontaneous emission obtained in Eqs. (54) and (55). The PL intensity as a function of $\hbar\omega$ has been measured in a-Si:H. From these measurements the photon energy corresponding to the PL peak maximum can be determined. By comparing the experimental energy thus obtained with the energy corresponding to the maximum of the rate of spontaneous emission, we can determine E_0. For this purpose, we need to determine the energy at which the rates in Eqs. (54) and (55) become maximum.

Defining $x = \hbar\omega - E_0$, $(x > 0)$ and $\beta = 1/\kappa_B T$ and then setting $dR_{spi}/dx = 0$ ($i = 1, 2$), we get x_{01} and x_{02} from Eq. (54) or Eq. (55) at which the rates are maximum, respectively, as:

$$x_{01} = \frac{1 - \beta E_0}{2\beta} \pm \sqrt{\frac{(1 - \beta E_0)^2}{4\beta^2} + \frac{2E_0}{\phi}}, \tag{56}$$

and

$$x_{02} = \frac{3 - \beta E_0}{2\beta} \pm \sqrt{\frac{(1 - \beta E_0)^2}{4\beta^2} + \frac{2E_0}{\phi}} \, , \qquad (57)$$

where only the + sign produces $x > 0$. It may be noted that both Eqs. (54) and (55) give the same expression for x_{01} and x_{02} as derived in Eqs. (56) and (57). Using $x_0 = E_{mx} - E_0$, where $E_{mx} = \hbar\omega_{max}$ is the emission energy at which the PL peak intensity is observed, the corresponding E_0 is obtained from Eqs. (56) and (57)as:

$$E_{01} = \frac{E_{mx}(-1 + \beta E_{mx})}{1 + \beta E_{mx}} \, , \qquad (58)$$

$$E_{02} = \frac{E_{mx}(-3 + \beta E_{mx})}{1 + \beta E_{mx}} \, . \qquad (59)$$

Using this in Eq. (58) and (59) one can calculate the rates of spontaneous emission from the two approaches for all the four possibilities. The time of radiative recombination or radiative lifetime, τ_{ri}, is then obtained from the inverse of the maximum rate ($\tau_{ri} = 1/R_{spi}$, i = 1, 2) obtained from Eqs. (54) and (55) calculated at $\hbar\omega = E_{mx}$. However, before we can calculate the rates of spontaneous emission due to excitonic recombination in a-semiconductors from Eqs. (54) and (55), we need to know the corresponding effective masses of charge carriers for calculating the excitonic reduced mass μ_x and Bohr radius a_{ex}.

A theory of calculating the effective mass of charge carriers in amorphous solids has recently been developed [22]. It is also briefly described below.

Effective Mass of Charge Carriers

The effective masses of e and h have recently been derived for amorphous solids [5, 20-23]. Accordingly, the effective mass of an electron in the conduction extended states is obtained as:

$$m_{ex}^* = \frac{E_L}{2(E_2 - E_c)a^{1/3}} m_e \, , \qquad (60)$$

where E_2 is the energy of the middle of the extended states at which the imaginary part of the dielectric constant becomes maximum [32] and m_e is the free electron mass. The effective mass of an electron in the conduction tail states is obtained as:

$$m_{et}^* = \frac{E_L}{2(E_c - E_{ct})b^{1/3}} m_e \, , \qquad (61)$$

where E_{ct} is the energy of the end of the conduction tail states. Likewise, expressions for the effective masses of a hole in the valence extended (m_{hx}^*) and tail (m_{ht}^*)states are respectively obtained as:

$$m_{hx}^* = \frac{E_L}{2(E_v - E_{v2})a^{1/3}} m_e \ ,$$

(62)

and

$$m_{ht}^* = \frac{E_L}{2(E_{vt} - E_v)b^{1/3}} m_e \ ,$$

(63)

where E_{v2} and E_{vt} are the energies corresponding to the half width of valence extended states and top of the valence tail states above the hole mobility edge. For sp^3 hybrid systems, we can estimate the energies E_c - E_{ct} and E_{vt} - E_v as E_c - E_{ct} = E_{vt} - E_v = E_c /2 [22]. The energy E_L is given by:

$$E_L = \frac{\hbar^2}{m_e L^2} \ .$$

(64)

In Eqs.(60) - (63), a and b are the fractional concentrations of atoms contributing to the extended and tail states, respectively. It is to be noted that for a sp^3 hybrid system, as the effective mass depends on the bandwidth, the effective mass of an electron is the same as that of a hole [22]. This is because the valence and conduction bands in these systems are the two halves of the overlapping sp- band and hence have the same bandwidth. Thus, for a-Si:H we get $m_e^* = m_h^*$.

Results

Rates of excitonic spontaneous emission are derived: (a) within two-level approximation, (b) under non-equilibrium with the joint and product density of states and (c) under equilibrium with the product density of states. Although, the rate derived within the two level approximation is strictly valid only for isolated atomic systems, for applying it to a condensed matter system one requires ω and $|r|$. For applying it to the excitonic radiative recombination, one may be able to assume $|r| = a_{ex}$, but it is not clear at what value of ω one should calculate the rate to determine the radiative lifetime. For this reason, the rate R_{sp12} in Eq. (32) derived within the two-level approximation will be calculated later at the end of this section.

Rates in Eqs. (46) -(48), derived under non-thermal equilibrium with the product density of states, cannot be used to calculate the radiative lifetime unless one knows the value of E_0, which cannot be determined from these expressions. Therefore, here we can only present the results of the rates of spontaneous emission derived under the thermal equilibrium in Eqs. (54) and (55). We will use Eq. (54) to calculate the rates

R_{spi}, $i = 1,2$, for the possibilities (i) to (iii) and Eq. (55) to calculate R_{spt1} and R_{spt2} for the possibility (iv) in a-Si:H. For this first we need to know the effective masses of charge carriers.

For calculating the effective masses of charge carriers, we consider a sample of a-Si:H with 1 at.% weak bonds contributing to the tail states (i.e. $a = 0.99$ and $b = 0.01$), using $L = 0.235$ nm [33], $E_2 = 3.6$ eV [34], $E_c = 1.80$ eV [35] and $E_c - E_{ct} = 0.8$ eV [36], we get from Eqs. (60) and (61) $m_{ex}^* = 0.34\ m_e$ and $m_{et}^* = 7.1\ m_e$, respectively, for a-Si:H.

For determining E_0 from Eqs. (58) and (59), we need to know the values of E_{mx} and the carrier (exciton) temperature T. Wilson et al. [6] have measured PL intensity as a function of the emission energy for three different samples of a-Si:H at 15 K and at two different decay times of 500 ps and 2.5 ns. Stearns [4] and Aoki [7,37] have also measured it at 20 K and 3.7 K, respectively. The PL spectra measured in these three experiments are reproduced in Fig. 2 and E_{mx} estimated from these spectra is given in Table 1.

Table 1 Values of E_{mx} estimated from the maximum of the observed PL intensity from three different experiments (Fig. 2) and the corresponding values of E_0 calculated from Eqs. (58) and (59). (Note that both expressions produce the same value.)

Experiment	T (K)	E_{mx} (eV)		E_0 (eV)	
		500 ps	2.5 ns	500 ps	2.5 ns
Sample 1[6]	15	1.428	1.401	1.425	1.398
Sample 2[6]	15	1.444	1.400	1.441	1.397
Sample 3[6]	15	1.448	1.405	1.445	1.402
Sample 4 [4]	20	1.450		1.447	
Sample 5 [37]	3.7	1.360		1.359	

The next problem is to find the temperature of the excited carriers before they recombine radiatively. This can also be done on the basis of the three experimental results on PL [4, 6, 37]. The measured energy, E_{mx}, of maximum PL intensity is below the mobility edge E_c by 0.4 eV [4,6] to 0.44 [37]. That means most excited charge carriers have relaxed down below their mobility edges and are not hot carriers anymore before the radiative recombination. It is therefore only logical to assume that the excited charge carriers in these experiments are in thermal equilibrium with the lattice and E_0 should be calculated at the lattice temperature. The assumption is very consistent with the established fact that the carrier-lattice interaction is much stronger in a-Si:H [22] than in crystalline Si.

Using the experimental values of E_{mx} and the corresponding lattice temperature, the values of E_{01} calculated from Eq. (58) and those of E_{02} calculated from Eq. (59) are also listed in Table 1. Both the approaches, Eq. (58) and Eq. (59), produce identical results for E_0. It should be noted, however, that E_0 increases slightly with the lattice temperature.

The calculated E_0, from the two time resolved spectra at two different decay times by Wilson et al. [6], is found to be slightly lower at 2.5 ns than at 500 ps because the corresponding observed E_{mx} shifts to a lower energy in 2.5 ns. The shifting of E_0 to a lower energy with time suggests a relaxation of the system to lower energy configurations, which agrees with the interpretation of the experimental results by Wilson et al. [6]. Having determined the effective mass and E_0, we can now calculate the rates of spontaneous emission for transitions through possibilities (i) - (iv).

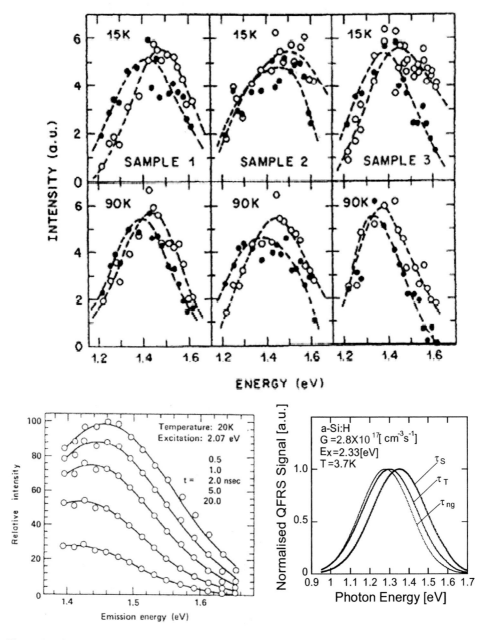

Figure 2: Time resolved PL spectra of (a) three samples of a-Si:H at 15 K [6], (b) at 20 K [4], and (c) QFRS PL spectra for the τ_S (72.3 kHz), τ_T (49.7 Hz) and τ_{ng} (0.280 Hz) lifetime peaks with $G \approx 2.8 \times 10^{17}\,\mathrm{cm^{-3}\,s^{-1}}$ [37]. (All peaks of spectra in (c) are normalised to unity.)

(i) Both e and h excited in extended states

In the possibility (i), where both e and h are in their respective extended states, the excitonic reduced mass is obtained as $\mu_x = 0.17\,m_e$. Then using $n = 4$, the calculated R_{sp1} is plotted as a function of the emission energy and at a temperature $T = 15$ K for one of the samples of a-Si:H of Wilson et al. [6] in Fig. 3(a) and R_{sp2} for the same sample in Fig. 3(b). Similar curves are obtained for all the samples at other temperatures as well but the magnitudes of the rates vary.

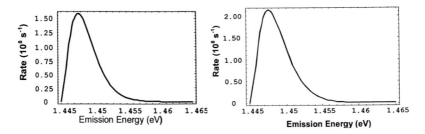

Figure 3: Rates of spontaneous emission plotted as a function of the emission energy for $E_0 =1.445$ eV at a temperature of 15 K and calculated from (a) method 1 and (b) method 2 for calculating the matrix elements.

Table 2: The maximum rates, R_{sp1} and R_{sp2}, of spontaneous emission and the corresponding radiative lifetime calculated using the values of E_{mx} and E_0 for the 5 samples in Table 1. Results are given for transitions involving extended – extended states [possibility (i)], extended-tail states [possibility (ii) and (iii)] and tail – tail states [possibility (iv)].

Sample	T (K)	R_{sp1} (s^{-1})	τ_{r1} (s)	R_{sp2} (s^{-1})	τ_{r2} (s)
1[a]	15	1.58×10^8	6.34×10^{-9}	2.05×10^8	4.88×10^{-9}
2[a]	15	1.56×10^8	6.42×10^{-9}	2.07×10^8	4.82×10^{-9}
3[a]	15	1.55×10^8	6.43×10^{-9}	2.00×10^8	5.00×10^{-9}
4[a]	20	2.77×10^8	3.62×10^{-9}	3.72×10^8	2.69×10^{-9}
5[a]	3.7	7.77×10^6	0.13×10^{-6}	8.38×10^6	0.12×10^{-6}
1[b]	15	2.97×10^8	3.37×10^{-9}	3.43×10^8	2.93×10^{-9}
2[b]	15	2.93×10^8	3.41×10^{-9}	3.46×10^8	2.89×10^{-9}
3[b]	15	2.92×10^8	3.42×10^{-9}	3.47×10^8	2.89×10^{-9}
4[b]	20	5.22×10^8	1.92×10^{-9}	1.08×10^9	0.92×10^{-9}
5[b]	3.7	1.47×10^7	0.68×10^{-7}	1.62×10^7	0.62×10^{-7}
1[c]	15	1.04×10^7	0.96×10^{-7}	1.18×10^7	0.85×10^{-7}
2[c]	15	1.03×10^7	0.97×10^{-7}	1.19×10^7	0.84×10^{-7}
3[c]	15	1.03×10^7	0.97×10^{-7}	1.19×10^7	0.84×10^{-7}
4[c]	20	1.84×10^7	0.55×10^{-7}	2.14×10^7	0.47×10^{-7}
5[c]	3.7	0.51×10^6	1.96×10^{-6}	0.55×10^6	1.82×10^{-6}

[a]Extended-extended states transitions (possibility (i))

[b] Extended- tail or tail-extended states transitions (possibilities (ii) – (iii))

[c] Tail-tail states transitions (possibility (iv))

The radiative lifetime can be obtained from the inverse of the maximum value of the rate at any temperature and E_0. The maximum value of the rates is obtained at the emission energy, E_{mx}, at any temperature. The rates and the corresponding radiative lifetimes thus calculated are listed in the Table 2. The maximum rate, R_{sp1}, at 15 K for the three samples of Wilson et al. [6] is obtained about 1.56×10^8 s^{-1} and the corresponding radiative lifetime $\tau_{r1} \approx 6.42$ ns, for the sample of Stearns [4] at 20 K $R_{sp1} = 2.77 \times 10^8$ s^{-1} and $\tau_{r1} = 3.6$ ns and for the sample of Aoki et al. [37] at 3.7 K $R_{sp1} = 7.77 \times 10^6$ s^{-1} and $\tau_{r1} = 0.13$ μs. For calculating the rate R_{sp2} ($i = 2$) in Eq. (54), we need to know a_{ex}, which is obtained from Eq. (49) as 4.67 nm in the extended states. The rates R_{sp2} thus obtained are slightly higher than the corresponding R_{sp1} at 15K, 20 K and 3.7 K (see Table -2). The corresponding radiative lifetime is in the ns time range at 15 K and 20 K and in the μs range at 3.7 K.

(ii)-(iii) Extended-to-tail states recombinations

For possibilities (ii) and (iii), where one of the charge carriers of an exciton is in its extended states and the other in its tail states, we get $\mu_x = 0.32$ m_e. Using this and other quantities same as in the possibility (i) above, we get from Eq. (54) ($i = 1$), $R_{sp1} = 2.94 \times 10^8$ s^{-1}, which gives $\tau_{r1} = 3.41$ ns at 15K, 5.22×10^8 s^{-1} giving $\tau_{r1} = 1.92$ ns at 20 K and at 3.7 K we get $R_{sp1} = 1.47 \times 10^7$ s^{-1}, which gives $\tau_{r1} = 0.1$ μs.

Likewise, for calculating R_{sp2} from Eq. (54) ($i = 2$) we need the corresponding excitonic Bohr radius, which is obtained from Eq. (49) as $a_{ex} = 2.5$ nm. Using this in Eq. (54), we get $R_{sp2} = 3.45 \times 10^8$ s^{-1} and $\tau_{r2} = 2.9$ ns at 15 K and $R_{sp2} = 1.08 \times 10^9$ s^{-1} giving $\tau_{r2} = 0.92$ ns at 20 K. At 3.7 K, we obtain $R_{sp2} = 1.62 \times 10^7$ s^{-1} and $\tau_{r2} = 0.1$ μs.

Thus, the radiative lifetimes for all the three possibilities (i) - (iii) are in the ns time range at 15K and 20 K, but in the μs time range at 3.7 K. However, the dependence of the rates on $x = \hbar\omega - E_0$ remains the same for all the possibilities, as shown in Figs. 3(a) and 3(b).

(iv) Tail-to-tail states recombinations

For the possibility (iv), when both e and h are localized in their tail states, we get $\mu_x = 3.55$ m_e, $a_{ex} = 0.223$ nm and $t'_e = 1.29 \times 10^{10}$ m^{-1}. Using these in Eq. (55), with $i = 1$, one gets $R_{spt1} = 1.03 \times 10^7$ s^{-1}, which gives $\tau_{r1} = 0.97 \times 10^{-7}$s at 15K, $R_{spt1} = 1.84 \times 10^7$ s^{-1}, which gives $\tau_{r1} = 0.55 \times 10^{-7}$ s at 20 K and $R_{spt1} = 0.51 \times 10^6$ s^{-1} with the corresponding $\tau_{r1} = 2.0$ μs at 3.7 K.

Using Eq. (55) with $i = 2$, we get $R_{spt2} = 1.18 \times 10^7$s^{-1} which gives $\tau_{r2} = 0.85 \times 10^{-7}$s at 15K, and $R_{spt2} = 2.14 \times 10^7$s^{-1}, which gives $\tau_{r2} = 0.47 \times 10^{-7}$ s at 20K. The corresponding results at 3.7 K are obtained as $R_{spt2} = 0.55 \times 10^6$ s^{-1} and $\tau_{r2} = 2.0$ μs. Thus, for tail-to-tail states transitions, the radiative lifetime is found to be much longer, in the μs time range. This agrees very well with the PL lifetimes measured in all the three experiments.

Results of two-level approximation

As we have determined the energy, E_{mx}, at which maximum of PL peaks have been observed in the five samples of a-Si:H, we can calculate R_{sp12} in Eq. (32) and the corresponding radiative lifetime at these photon energies for a comparison. Here we will use $|r| = a_{ex}$ and $\omega = E_{mx}/\hbar$ in Eq. (32). For extended-to-extended state recombination, $a_{ex} = 4.67$ nm and using $E_{mx} = 1.4$ eV for Wilson et al.'s sample, we get $\omega = 2.1 \times 10^{15}$ Hz, $R_{sp12} = 7.4 \times 10^{10}$ s^{-1} and $\tau_r = 1/R_{sp12} = 13$ ps. For Stern's sample, with $E_{mx} = 1.45$ eV leading to $\omega = 2.2 \times 10^{15}$ Hz, we get $R_{sp12} = 8.5 \times 10^{10}$ s^{-1} and the corresponding $\tau_r = 12$ ps. For Aoki's sample with $E_{mx} = 1.36$ eV and the corresponding $\omega = 2.0 \times 10^{15}$ Hz, we get $R_{sp12} = 6.2 \times 10^{10}$ s^{-1} and $\tau_r = 16$ ps. Thus, within the two level approximation, one cannot get any results for the radiative lifetime longer than ps, which does not agree with the experimental results. This is a further clear indication of the fact that the two-level-approximation is not valid for condensed matter systems.

Discussions

A comprehensive theory for calculating the radiative lifetime of excitons in amorphous semiconductors is presented. Using the electron-photon interaction Hamiltonian as a perturbation, the rate of spontaneous emission is calculated by applying Fermi's golden rule under several possible conditions. The radiative lifetime is then obtained from the inverse of the maximum rate thus calculated. The theory is applied to calculate the rate of spontaneous emission for five samples of a-Si:H for which the PL spectra have been measured at 15 K [6], 20 K [4] and 3.7 K [37]. The rate increases and hence the radiative lifetime decreases with the PL energy, which is quite consistent with the observed results shown in Fig. 1 for τ_2. (Note that τ in Fig. 1 is plotted on the logarithmic scale). The emission energy E_{mx}, at which the measured PL intensity is maximum, changes with the temperature and time-delay (see Fig. 2). Assuming that the theoretical spontaneous rate of emission is maximum at the same energy as E_{mx} and at the same temperature, the energy E_0 is determined from Eqs. (58) and (59), which produces different values of E_0 at different values of E_{mx}. Such a change in E_0 can only be possible in amorphous solids, which do not have a well defined energy gap and therefore the excited charge carriers are able to relax down to different energy levels, E_0, through the four different possibilities.

In the previous section we calculated the radiative lifetime and found that for the possibilities (i) – (iii), the radiative lifetime is in the ns time range at temperatures > 15 K, but in the μs time range at a lower temperature (3.7 K). For the possibility (iv), the radiative lifetimes are much longer than those for the possibilities (i) – (iii). It is not possible from these results to determine where exactly the PL is originating from in these experiments, i.e. is it from the extended or tail states? In order to determine the origin of PL, one also needs to know the Stokes shift in the PL spectra. For example, if the excitonic PL is occurring through radiative transitions from extended-to-extended states, then the observed Stokes shift observed in the PL spectra should only be equal to the exciton binding energy. For this reason it is important to determine the exciton binding energy corresponding to all the four transition possibilities.

The ground state singlet exciton binding energy E_s in a-Si:H is obtained as [5,22]:

$$E_s = \frac{9\mu_{ex}e^4}{20(4\pi\varepsilon_0)^2\varepsilon^2\hbar^2} \ . \tag{65}$$

This gives $E_s \sim 16$ meV for the possibility (i), 47 meV for the possibilities (ii) and (iii) and 0.33 eV for the possibility (iv). The known optical gap for a-Si:H is 1.8 eV [35] and E_0 estimated from experiments is about 1.44 eV measured at 15 K [6], 1.45 eV measured at 20 K [4], and 1.36 eV measured at 3.7 K [37] (see Table 1). Considering that the PL in a-Si:H originates from excitonic states [possibility (i)], we find a Stokes shift of 0.36 eV, 0.35 eV and 0.44 eV at temperatures 15K, 20 K and 3.7 K, respectively. Such a large Stokes shift is not possible only due to the exciton binding energy, which at most is in the meV range for possibilities (i) - (iii) as given above. As the non-radiative relaxation of excitons is very fast, not much PL may occur through the possibility (i) even at a time delay of 500 ps measured by Wilson et al. [6] and Stearns [4]. It may be possible to occur from transitions through the possibilities (ii) - (iv), which means at least one of the charge carriers has relaxed to the tail states before recombining radiatively. This can be easily explained from the following: As stated above, the charge carrier-lattice interaction is much stronger in a-Si:H than in crystalline Si [22]. As a result, it is well established [33] that an excited hole gets self-trapped very fast in the tail states in a-Si:H. Thus a Stokes shift of about 0.4 eV observed experimentally at 3.7 K, 15 K and 20 K is due to the relaxation of holes in excitons to the tail states plus the excitonic binding energy, which is in the meV range and hence plays very insignificant role. The radiative lifetimes of such transitions calculated from the present theory and listed in Table 2 fall in the ns time range at temperatures 15 – 20 K which agrees very well with those observed by Wilson et al. [3, 6] and Stearns [4]. At 3.7 K, the calculated radiative lifetime is found to be in the μs range, which also agrees very well with Aoki et al.'s experimental results of the singlet exciton's radiative lifetime [37] (see Figs. 1 and 4). QFRS spectra of a-Si:H and a-Ge:H measured at various temperatures by Aoki et al. [37] are reproduced here in Fig. 4.

Wilson et al. [6] have also measured PL at a longer time delay of 2.5 ns and the value of E_{mx} determined from that measurement is 1.44 eV, lower than 1.45 eV observed at a shorter time delay of 500 ps. This is a clear indication that the excited charge carriers have relaxed in energy by 10 meV in 2.5 ns. They have also observed the lifetime in the μs time range range, which may be attributed to transitions through the possibility (iv). Aoki et al. [37] have observed another PL peak at 3.7 K (τ_g, Fig 4) with a radiative lifetime in the μs range but longer than the singlet exciton radiative lifetime (τ_2) at 3.7 K. According to the radiative lifetime calculated here (table 2), the PL peak at τ_g may be attributed to transitions from the tail-to-tail states (possibility (iv)) because the corresponding radiative time is about 2 μs, longer than the radiative lifetime for possibilities (i) – (iii).

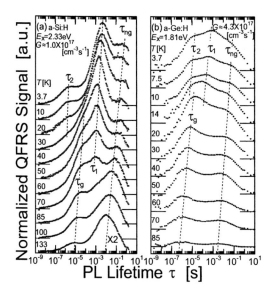

Figure 4: QFRS spectra of a-Si:H and a-Ge:H photoexcited at energy , E_{ex} = 2.33 eV, with G = 4.3 x 10^{17}cm^{-3}s^{-1} [37] observed at various temperatures.

For the possibility (iv), when both the excited charge carriers are in their tail states, which are localized states, the concept of excitons is not valid and such an excited pair of charge carriers is called a geminate pair instead of an exciton. In this case the excitonic wave function cannot be used to evaluate the integral $Z_{lm\lambda}$, one has to use the wave functions of e and h [see Eq. (10)]. According to the present approach, when an exciton formed initially through the possibility (i), relaxes down to the possibilities (ii) - (iv), its energy and excitonic Bohr radius are restricted by the quantum rules of excitonic states. Thus, in the possibility (iv) although the charge carriers form a geminate pair, their separation will still be the same as the excitonic Bohr radius before their radiative recombination. In this view, there are two types of geminate pairs possible in amorphous semiconductors, one from the relaxed exciton to the tail states and the other excited e and h pair relaxed to the tail states. In the second case the separation between e and h may be larger than the excitonic Bohr radius and such pairs are also referred to as non-geminate pairs [7]. Indeed, in Fig. 4 a peak, denoted by τ_{ng}, appears at much larger lifetime of ms and Aoki etal. have attributed this peak to the non-geminate pairs. According to the present theory, this peak may be attributed to the second kind of geminate pairs, which relax down to the tail states without forming excitons and hence are the same as non-geminate pairs.

The longer lifetime obtained for the possibility (iv) is because of the exponential factor $\exp(-2t'_e a_{ex})$ appearing in the rate of recombination in Eq.(55) through Eq.(48), which is proportional to the probability of quantum tunnelling a barrier of height E_c for a distance a_{ex}. Therefore, the expression also implies that the rate of spontaneous emission reduces and hence the radiative lifetime gets prolonged in the possibility (iv) due to the quantum tunnelling through a distance of a_{ex} before the radiative recombination.

In evaluating the transition matrix element of the exciton-photon interaction operator [Eq. (3)], which only depends on the relative momentum between e and h in an exciton, the wave functions of e and h are used instead the wave function of an exciton [5], for example, in the form of $\Psi(R,r) = \exp\left(i\dfrac{\mathbf{P}\cdot\mathbf{R}}{\hbar}\right)\phi(r)$. From this point of view, the theory presented here appears to address the recombination between an excited pair of e and h not that between e and h in an exciton. The fact that the interaction operator is independent of the centre of mass momentum $\mathbf{P}$, the integral $Z_{lm\lambda}$ [Eq. (10)] always gives the average value of the separation between e and h. In an exciton, the separation is the excitonic Bohr radius and in an e and h pair it would be their average separation. As far as the recombination in amorphous solids is concerned, therefore, this is the only difference between an excitonic recombination and free e and h recombination. For excitons, using the excitonic wave function also, one would get the same value of the integral for possibilities (i) - (iii). However, for the possibility (iv), the exciton wave functions do not give the correct result, which indicates that the excitonic concept is not applicable when the excited charge carriers have relaxed to the tail states, as explained above.

The two approaches used here for calculating the transition matrix element produce similar results for the rate of radiative recombination, although their dependence on the emission energy is different. Such a difference is well known in calculating the absorption coefficient, where the approach 1 gives Tauc's relation and approach 2 accounts for the deviations observed from Tauc's relation [21,22].

Temperature Dependence

The maximum rates calculated from Eqs. (54) and (55) at the emission energy E_{mx} increase exponentially as the temperature increases. This agrees very well with the model used by Wilson et al. [6] to interpret the observed temperature dependence in their PL intensity. Such a temperature dependence is independent of the origin of PL, whether the recombination originates from the extended or tail states. However, the maximum rates obtained in Eqs. (54) and (55) vanish at $T = 0$ because x_{01} [Eq.(56)] and x_{02} [Eq. (57)] become zeros as $T \to 0$, which is in contradiction with the observed rates [4, 6]. The reason for this is that one cannot derive the rate at 0 K from Eqs. (54) and (55). At 0 K the, the electron and hole probability distribution functions become unity; $f_c \to 1$ for $E_e < E_{Fn}$ and $f_v \to 1$ for $E_h > E_{Fp}$ as $T \to 0$ and using this the rates at 0 K become equal to the pre-exponential factors as obtained in Eqs. (46) - (48). Thus, the temperature dependence of the rates obtained here can be expressed as:

$$R_{spi}(T) = R_{oi}[(1 - \Theta_1(T)) + \exp[-(\hbar\omega - E_0)/\kappa_B T]\Theta_2(\hbar\omega - E_0)], \; i = 1, 2 , \quad (66)$$

where R_{0i} are the pre-exponential factors of Eqs. (54) and (55) corresponding to the four possibilities. The first step function, $\Theta_1(T)$, is used to indicate that the first term of Eq. (66) vanishes for $T > 0$ and the second step function $\Theta_2(T)$ represents that there is no spontaneous emission for $\hbar\omega - E_0 < 0$. For an estimate of R_{0i} we have used the values of E_{mx} and E_0 obtained at $T = 3.7$ K, which gives $R_{01} = 7.77 \times 10^6$ s^{-1}, 1.47×10^7 s^{-1} and 0.51×10^6 s^{-1} for extended -extended, extended-tail and tail-tail states, respectively, and the corresponding values for R_{02} are obtained as 8.38×10^6 s^{-1}, 1.62×10^7s^{-1} and 0.55×10^6s^{-1}.

Wilson et al. [6] have fitted their observed rates to the following model:

$$\frac{1}{\tau_r} = v_1 + v_0 \exp(T/T_0), \qquad (67)$$

and the best fit has been obtained with $v_1 = 10^8 \mathrm{s}^{-1}$ and $v_0 = 0.27 \times 10^6$ s^{-1} and $T_0 = 95$ K. Stearns [4] has obtained best fit to his data with $v_1 = 2.7 \times 10^8$ s^{-1} and $v_0 = 6.0 \times 10^6$ s^{-1} and $T_0 = 24.5$ K. According to this model $\tau_r^{-1} = v_1$ at $T = 0$. Comparing these with the present theory, one should have $R_{0i} = v_1$, R_{0i} is an order of magnitude smaller than v_1. Also in Eq. (67), $v_1 > v_0$ whereas according to our results in Eq. (66) $v_1 = v_0 = R_{0i}$. These discrepancies may be attributed to the fact that Eq. (67) is obtained by fitting to the experimental data and not by any rigorous theory. As a result, both Eqs. (66) and (67) produce similar results for the radiative lifetime, but individual terms on the right hand sides contribute differently. This is also apparent from the fact that according to Eq. (67), v_1 contributes to the rates also at non-zero temperatures but according to the derived rate in Eq. (66), R_{0i} contributes only at $T = 0$ K. It may also be noted that the temperature dependence of the rate of spontaneous emission in Eq. (67) fitted to their experimental data by Wilson et al. is different from their observed PL intensity dependence on temperature and it is also different from the temperature dependence derived here in Eq. (66). Theoretically, by assuming that the rate is maximum at the same energy as the PL peak, we get only one kind of dependence on temperature for both, the rates of spontaneous emission and PL intensity. Stearns [4] has fitted the observed rate of emission at the emission energy of 1.43 eV to Eq. (67). We have plotted the calculated rates in Eq. (66) as a function of the temperature for the possibility (ii) at the same emission energy (1.43 eV) and shown in Figs. 5(a) and (b), respectively.

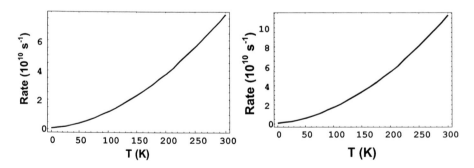

Figure 5: Rates plotted as a function of temperature at an emission energy of 1.43 eV and calculated for the possibility (ii) from Eq. (54) in (a) $i = 1$ and (b) $i = 2$.

Recently Aoki et al. [37] have measured the QFRS spectra of a-Si:H and a-Ge:H at various temperatures as shown in Fig. 4, where all PL peaks appearing at $\tau_2 (\mu\,\mathrm{s})$, $\tau_g (\mu\,\mathrm{s})$, $\tau_1 (\mathrm{ms})$ and $\tau_2 (0.1\,\mathrm{s})$ show a shift toward the shorter time scale as the temperature increases. This behaviour is quite consistent with the temperature dependence obtained from the present theory.

Excitonic Concept

Another question may arise here that why is the concept of excitons necessary to explain the radiative lifetime when the excited charge carriers can recombine without forming excitons? This can be answered as follows: Instead of using the reduced mass of excitons and the excitonic Bohr radius, if one uses the effective masses of e and h and the distance between their separation, one can calculate the radiative lifetime of a free excited electron-hole pair (not an exciton) from Eqs. (54) and (55). The radiative lifetime thus obtained for such a recombination is also in the ns range for the possibilities (i) - (iii). For the possibility (iv), of course, as the separation between e and h is expected to be larger one may get smaller rates and hence longer radiative lifetime. However, there are two important observations which support the formation of excitons. 1) A free e and h pair will relax down to the tail states in a ps time scale resulting into a PL only from the possibility (iv), which is not supported by the observed PL (see Fig. 4). 2) PL has a double peak structure appearing in the time resolved spectra corresponding to the singlet and triplet excitons. No such spin correlated peaks can be observed with the recombination of free electron and hole pairs.

The theory presented here for the radiative recombination from the tail states is different from the radiative tunnelling model [35], where the rate of transition is given by: $R_{spt} \sim R_0 \exp(-2d/d_0)$, d being the separation between e and h, d_0 is the larger of the extents of the electron and hole wave functions, usually considered to be 10-12 Å and R_0 is the limiting radiative rate expected to be about $10^8 - 10^9$ s^{-1}. Here d, being the distance between a pair of excited charge carrier (not an exciton), is not a fixed distance, like a_{ex}, it can be of any value. As a result, the radiative tunnelling model breaks down not only at early times after the excitation where the recombination rates are of the same order as the limiting rate as it has been already pointed out by Wilson et al. [3], but it also cannot explain the appearance of the double peak structure in the PL originating from the tail states. Although the rate of recombination from the tail states derived in Eq. (55) also has a similar exponential dependence as in the radiative tunnelling model, it depends on the excitonic Bohr radius which is a constant and hence the exponential factor becomes a constant. This explains very well the peak structure in the PL.

Kivelson and Gelatt [2] have developed a comprehensive theory for PL in amorphous semiconductors based on a trapped exciton model. A trapped exciton considered by them is in which the hole is trapped in a localized gap state and the electron is bound to the hole by their mutual Coulomb interaction. The rate of radiative transition is calculated by applying Fermi's golden rule from an initial state of an exciton bound in a hydrogenic state with a large radius a_B^* to a final state of a trapped hole with a much smaller radius a_h. The model is thus similar to the possibility (ii) considered in this paper, but the possibility (i) has not been considered by them. The square of the dipole transition matrix element is estimated to be $\sim (2a_h)^5 / (a_B)^3$ by them. This is probably one of the main differences between the present theory and the theory of Kivelson and Gelatt. Here we have presented two approaches for calculating the transition matrix element. In the first approach, the transition matrix element is assumed to be a constant, Z_1 [24] as given in Eq. (13). In the second approach, it is equal to Z_2, derived from the dipole approximation as given in Eq. (14). Although the rates of spontaneous emission calculated from the two approaches are nearly the same and have the same temperature dependence, their expressions have different PL photon energy dependence. The difference in the photon energy dependence is also found in the absorption coefficient, which is well known [21,22].

The other important difference between the present theory and theory of Kivelson and Gelatt is in the possibility (iv) involving recombinations in the tail states. For this possibility, Kivelson and Gelatt have followed the radiative tunnelling model and used the exponential factor as $\exp(-2d/a_B^*)$. As discussed above such an exponential dependence on varying d cannot explain the peak structures observed in the PL spectra.

It is quite clear from the expressions derived in Eqs. (54) and (55) that rates do not depend on the excitation density. This agrees very well with the measured radiative lifetime by Aoki et al. [37], who have found that the radiative lifetime of singlet and triplet excitons is independent of the generation rate. This provides additional support for the PL observed in a-Si:H to be due to excitons. One of the speculations, as described above, has been that the shorter PL time in the ns range observed by Wilson et al. may be due to the high excitation density used in the TRS measurements. The present theory clarifies this point very successfully that it is not the excitation density but it is the combined effect of temperature and fast non-radiative relaxation of charge carriers to lower energy states that one does not observe any peak in the ns time range at 3.7 K.

Considering distant pair recombinations, Levin et al. [19] have studied PL in amorphous silicon at low temperatures by computer simulation. They have also used the radiative tunnelling model for recombinations in the tail states, but they have not considered radiative recombination of excitons. Therefore, results of their work cannot explain the occurrence of the double peak structure in PL.

As stated above, the charge carrier-lattice interaction is known to be much stronger in a-Si:H and therefore its effect on PL should also be studied. Such a study will be the objective of future work, however, as the non-radiative relaxation is much faster (in ps) than the radiative recombination, it is expected that one may not have any measure influence on the other. When both radiative and non-radiative processes take place in the same time scale then their influence on each other is expected to be greater and more complicated.

Conclusions

In conclusion, it may be stated that the excitation density independent PL observed in a-semiconductors arises from the radiative recombination of excitons. Both the Stokes shift and radiative lifetimes should be taken into account in determining the PL electronic states. Although the radiative lifetime for possibilities (i) to (iii) are of the same order of magnitude, the Stokes shift observed in the PL suggest that these recombinations occur from extended-to-tail states [(possibility (ii)] in a-Si:H. The singlet radiative lifetime is found to be in the ns time range at temperatures > 15 K and in the μ s at 3.7 K, and triplet lifetime in the ms range.

In the possibility (iv), as carriers have to tunnel to a distance equal to the excitonic Bohr radius, the radiative lifetime gets prolonged. There are two types of It may also be pointed out that what is usually referred to as geminate recombination is nothing but the recombination between an e and h in an exciton that has relaxed to the tail states. The effective mass of a charge carrier changes as it crosses its mobility edge. This also influences the radiative lifetime of an exciton as it crosses the mobility edges. A large Stokes shift implies a strong carrier-lattice interaction in a-Si:H and therefore PL occurs in thermal equilibrium. Results of two-band approximation and non-equilibrium are therefore not applicable for a-Si:H.

Acknowledgements

We have very much benefited from discussions with Professors T. Aoki and K. Shimakawa during the course of this work. We are grateful to Professor T. Aoki also for allowing the reproduction of Figs. 2(c) prior to its publication. The work is supported by the Australian Research Council's large grants(2000-2003) and IREX (2001-2003) schemes.

References

[1] R.A. Street, *Adv. Phys.* **30**, 593 (1981).
[2] S. Kivelson and C.D. Gelatt Jr., *Phys. Rev.* B**26**, 4646 (1982).
[3] B.A. Wilson, P. Hu, T.M. Jedju and J.P. Harbison, *Phys. Rev.* B**28**, 5901 (1983).
[4] D.G. Stearns, *Phys. Rev.* B**30**, 6000 (1984).
[5] J. Singh, T. Aoki and K. Shimakawa, *Phil. Mag.* B**82**, 855 (2002).
[6] B.A. Wilson, P. Hu, J.P. Harbison and T.M. Jedju, *Phys Rev Letters* **50**, 1490 (1983).
[7] T. Aoki, S. Komedoori, S. Kobayashi, C. Fujihashi, A. Ganjoo and K. Shimakawa, *J. Non-Cryst. Solids* **299**-302, 642 (2002). The group has recently notified that an error was made in measuring the E_{mx} in this paper and that it has been corrected in ref. [27].
[8] S.P. Deppina and D.J. Dunstan, *Phil. Mag.* B**50**, 579 (1984).
[9] F. Boulitrop and D.J. Dunstan, *J. Non-Cryst. Solids* **77**-78, 663 (1985).
[10] R. Stachowitz, M. Schubert and W. Fuhs, *J. Non-Cryst. Solids* **227**-230, 190 (1998).
[11] S P. Deppina and D.J. Dunstan, *Phil. Mag.* B**50**, 579 (1984).
[12] F. Boulitrop and D.J. Dunstan, *J. Non-Cryst. Solids* **77**-78, 663 (1985).
[13] S. Ishii, M. Kurihara, T. Aoki, K. Shimakawa and J. Singh, *J. Non-Cryst. Solids* **266**-269, 721 (1999).
[14] T. Aoki, S. Komedoori, S. Kobayashi, T. Shimizu, A. Ganjoo and K. Shimakawa, *Nonlinear Optics* **29**, 273 (2002).
[15] T. Aoki, J. Mat. Sci. - *Materials in Engineering* **14**,697 (2003).
[16] W. Dumke, *Phys. Rev.* **105**, 139 (1957).
[17] G. Lasher and F. Stern, *Phys. Rev.* A**133**, 553 (1964).
[18] H. Barry Beb and E.W. Williams, in *Semiconductors and Semimetals*, Eds. R.K. Willardson and A.C. Beer (Academic, London, 1972), vol. 8, p. 18.
[19] E.I. Levin, S. Marianer, and B.I. Shklovskii, *Phys. Rev.* B**45**, 5906 (1992).
[20] J. Singh, J. Non-Cryst. *Solids* **299**-302, 444 (2002).
[21] J. Singh, *Nonlinear Optics* **29**, 111 (2002).
[22] J. Singh and K. Shimakawa, *Advances in Amorphous Semiconductors* (Taylor and Francis, London, 2003).
[23] J. Singh, *J. Materials Sci.* **14**, 171 (2003).
[24] N.F. Mott and E.A. Davis, *Electronic Processes in Non-crystalline Materials* (Clarendon Press, Oxford, 1979).

[25] G.D. Cody, in *Semiconductors and Semimetals*, vol. 21, part B (1984) p.11.

[26] P.K. Basu, *Theory of Optical Processes in Semiconductors* (Clarendon Press, Oxford, 1997).

[27] H.T. Grahn, *Introduction to Semiconductor Physics* (World Scientific, Singapore, 1999).

[28] J. Singh, *Excitation Energy Transfer Processes in Condensed Matter*(Plenum, N.Y., 1994).

[29] W. Van Roosbroeck and W. Shockley, *Phys Rev* **94**, 1558(1954).

[30] S.R. Elliott, *The Physics and Chemistry of Solids* (John Wiley & Sons, Sussex, 1998).

[31] J. Shah and R.C.C. Leite, *Phys. Rev. Lett.* **22**, 1304 (1969).

[32] S. Kivelson and C.D. Gelatt Jr., *Phys. Rev.* **B19**, 5160 (1979).

[33] K. Morigaki, *Physics of Amorphous Semiconductors* (World Scientific, London, 1999).

[34] L. Ley, in*: The Physics of Hydrogenated Amorphous Silicon II*, Eds. J.D. Joanpoulos and G. Lucovsky (Springer-Verlag, Berlin, 1984) P. 61.

[35] R.A. Street, *Hydrogenated Amorphous Silicon* (Cambridge Uni Press, Cambridge, 1991).

[36] W.E. Spear,W.E., in *Amorphous Silicon and Related Materials*, edited by H. Fritzsche (World Scientific, Singapore, 1988).

[37] T.Aoki, T. Shimizu, S. Komedoori, S. Kobayashi and K. Shimakawa, *J. Non- Cryst. Sol.* **338**-340, 456 (2004), and T. Aoki, T. Shimizu, D. Saito, K. Ikeda, Presented at 13 ISCMP, 1-3 September 2004, Varna, Bulgaria, published in J. Optoelectr. *Adv. Materials* **7**, 137(2005).

In: Trends in Materials Science Research
Editor: B.M. Caruta, pp. 147-156

ISBN: 1-59454-367-4
© 2006 Nova Science Publishers, Inc.

Chapter 6

The Thermoluminescent (TL) Properties of the Perovskite-Like KMgF₃ Activated by Various Dopants: A Review

C. Furetta and C. Sanipoli

Physics Department, Rome University "La Sapienza",
Piazzale A.Moro 2, 00185 Rome, Italy

Abstract

Thermoluminescence (TL) dosimetry has been developed to the stage that it represents a key technique in absorbed dose determination. TL dosimetry has found a very important use in clinical, personal and environmental monitoring of ionizing radiation. Interest in radiation dosimetry by the TL technique has resulted in numerous efforts seeking production of new, high performance TL materials. Of the many materials that have been produced and studied, several are now commonly used as thermoluminescent dosimeters (TLD). The aim of this paper is to present a review concerning the TL dosimetric characteristics as well as the kinetics parameters of the perovskite-like compound KMgF₃ in combination with various dopants. KMgF₃ has been growth and extensively studied since 1990 in the Physics Department of Rome University "La Sapienza" and the reported thermoluminescence characteristics have shown this phosphor to be a very good candidate for ionizing radiation dosimetry.

Introduction

Thermoluminescence (TL) dosimetry has been developed to the stage that it represents a key technique in absorbed dose determination. TL dosimetry has found a very important use in clinical, personal and environmental monitoring of ionizing radiation.

Interest in radiation dosimetry by the TL technique has resulted in numerous efforts seeking production of new, high performance TL materials. Of the many materials that have been produced and studied, several are now commonly used as thermoluminescent dosimeters (TLD).

The aim of this paper is to present a review concerning the TL dosimetric characteristics as well as the kinetics parameters of the perovskite-like compound $KMgF_3$ in combination with various dopants. $KMgF_3$ has been growth and extensively studied since 1990 in the Physics Department of Rome University "La Sapienza" and the reported thermoluminescence characteristics have shown this phosphor to be a very good candidate for ionizing radiation dosimetry.

$KMgF_3$ is a ternary compound belonging to the group of fluoroperovskites which have the general formula ABF_3, where A and B have the respective meanings alkali metal and alkaline earth metal. The $KMgF_3$ crystal has a typical cubic symmetry. When the basic material is doped with impurities, monovalent cations replace K^+ ions and anions, i.e. OH^-, substitute for F^- ions. If divalent or trivalent cations are used as dopants, they can replace Mg^{+2} or they are forced to occupy K^+ ion locations. The crystal cubic lattice of $KMgF_3$ is shown in **Fig.1**, where arrows indicate the possible positions of substitutional ions.

The interest in this particular material is caused by its very low hygroscopicity, which is of great importance to long-term use in different environmental conditions, its relatively high melting point, 1343 K, which allows high temperature annealing procedures. Furthermore, its effective atomic number Z_{eff} is 13 which is an intermediate value between tissue equivalent and high atomic number phosphors.

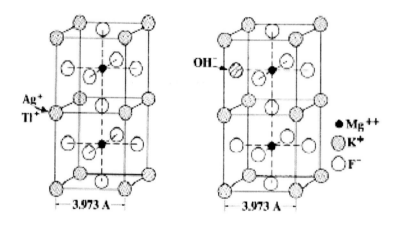

Fig.1. The crystal cubic lattice of $KMgF_3$

Material Preparation Method

Samples of perovskite $KMgF_3$ have been obtained from the melt using the kyropoulos technique. The molten mass was formed by heating at 1335 K, in a platinum crucible and under nitrogen atmosphere, the starting powder consisting of a finely ground stoichiometric mixture (mole ratio 1:1) of pure and dried KF and MgF_2. The growth of crystals was obtained with the pulling method from the above melt, starting the process with an air-cooled platinum finger or with the aid of a crystal seed. In the former case polycristalline ingots were obtained, in the latter single crystals were grown with typical dimensions of 2-3 cm in diameter and 2-3 cm in length. All samples were optically transparent. Doped crystals were obtained with the same technique by adding a proper amount of the desired inpurity to the melt.

The phosphors produced along the years were KMgF$_3$ doped with Pb and Eu (1990), KMgF$_3$:Tl (1994), KMgF$_3$ doped with Pb, Cr or Ag (1996) and Ce (1997/1999) and finally, in 2001/2002, doped with Er, La and Lu.

Because segregation and evaporation phenomena of the impurities occurred during the growth, the final dopand concentration in the crystal boule was not uniform. To obviate this problem, causing a large variation in the measurements carried out on different parts of the same ingot, in some cases the samples were reduced in powder and then mixed with PTFE for getting solid chips.

Dosimetric Properties of the Phosphors

-    pure KMgF$_3$ [1]

The thermoluminescent emission of pure KMgF$_3$ samples is shown in **Fig.2**; the glow curves have been recorded with heating rates of 5 and 29 Ks^{-1} respectively. The TL emission consists in a very intense peak and in a shoulder at higher temperature. The main peak can be attributed to the thermal activation of F centers created by irradiation in the material.

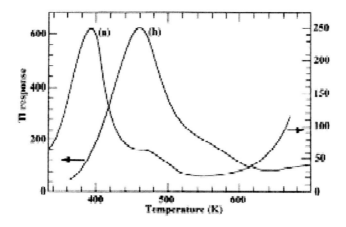

Fig.2. Glow curve of pure KMgF$_3$, beta irradiated. H.R. of 5 Ks^{-1} (a) and 20 Ks^{-1} (b)

-    Pb^{2+}-doped KMgF$_3$ [1,7]

Fig.3 shows the glow curve of this phosphor. The impurity content was 2 mol%. The TL emission is quite complex. The low temperature peaks correspond to the undoped material whereas the high temperature peaks are createdby the impurity ions. The TL intensity of KMgF$_3$:Pb is about two orders of magnitude smaller than that of pure KMgF$_3$. The increase of the dopant concentration does not increase appreciably the intensity.The high temperature peaks should arise from defects originated by association of the impurity ions with F centers. Indeed, Pb^{2+} ions are expected to replace substitutionally Mg2+ ions in the lattice, causing then two non-equivalent situations for the F centers.

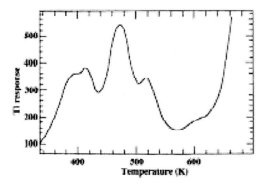

Fig.3. Glow curve of KMgF$_3$:Pb (2 mol%), beta irradiated. H.R. of 5 Ks^{-1}

Other samples of the same phosphor were obtained using an amount of impurity of only 0.5 mol% The glow curve is shown in **Fig.4**. Its shape is quite different from the previous one. Curve (b) has been obtained after a post-irradiation annealing at 50°C for 30 min; the h.r.used was 5°Cs^{-1}. The TL – dose response shows linearity up to 100 Gy.

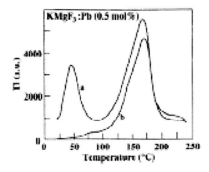

Fig.4. Glow curve of KMgF$_3$:Pb (o.5 mol%), beta irradiated. (a) pre-irradiation annealing at 400°C for 1 hr; (b) pre-irradiation annealing at 4000°C for 1 hr plus post-irradiation annealing at 50°C for 30 min.

-    <u>Eu^{3+}- doped KMgF$_3$ [1,2,3,4]</u>

KMgF$_3$:Eu exhibits intense thermoluminescence signals as shown in **Fig.5**. The glow curve consists of three well separated peaks. The first two maxima are roughly coincident with those in pure material; the third peak, caused by the impurity (0.2 mol%), is very prominent and can be used for dosimetric purposes.

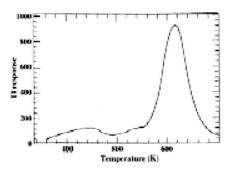

Fig.5. Glow curve of KMgF$_3$:Eu (0.2 mol%), betta irradiated. H.R. of 20 Ks^{-1}.

- ### TL$^+$-doped KMgF₃ [5,6]

KMgF₃:Tl single crystals were grown by the Kiropoulos method from a melt containing KF and MgF₂ in the stoichiometric ratio and about 1 mol% of the impurity. **Fig.6** shows a typical glow curve of this phosphor, obtained with a heating rate of 5 Ks^{-1}. The dose dependence of the TL emission shows a good linearity up to 100 Gy.

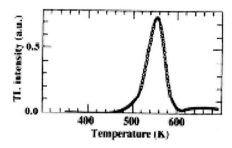

Fig.6. Glow curve of KMgF₃:Tl, beta irradiated. H.R. at 5 Ks^{-1}.

- ### Cr- and Ag-doped KMgF₃ [7]

Doping of KMgF₃ with Cr and Ag was obtained using CrF₃ (5 mol%) and AgI (2.5 mol%). The glow curves of these two phosphors are shown in **Fig 7** and the dose response in **Fig.8**.

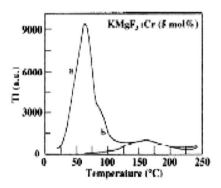

Fig.7. Glow curve of KMgF₃:Cr (5 mol%). Same annealing procedure as in Fig.4.

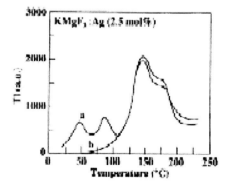

Fig.8. Glow curve of KMgF₃:Ag (25 mol%). Same annealing procedure as in Fig.4.

-    Ce- and Er doped KMgF$_3$ [8,9,10]

The glow curves of these two phosphors are very similar. As an example the glow curve of KMgF$_3$:Ce (0.5 mol%) is given in **Fig.9**. The glow curve structure changes drastically as the dopant concentration increases, as it is shown in **Fig.10** where the glow curve of KMgF$_3$ with 660 mol% of Ce is reported. A good linearity in the TL-dose response is obtained in the range 10^{-5} to 10^3 Gy fro both phosphors.

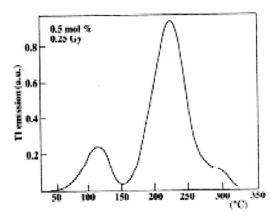

Fig.9 Glow curve of KMgF$_3$:Ce (0.5 mol%) after annealing at 550°C during 1 hr and a beta dose of 0.25 Gy.

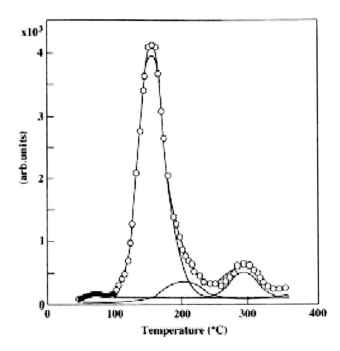

Fig.10. Glow curve of KMgF$_3$:Ce (660 mol%). The open circles are related to the experimental data and the solid line to the fitted curve

- <u>La-doped KMgF₃</u> [11,12]

A typical glow curve of this phosphor, having a dopant concentration of 2mol% is shown in **Fig.11**. This phosphor presents a good precison in the dose determination, 4.4%, and a good reproducibility, within 5%, of the TL signal over five repeated cycles of measurements carried out on the same sample. The stability of the TL signal, tested over a period of one month in dark condition, was very good: only 10% of the initial TL was lost over the whole storage time.

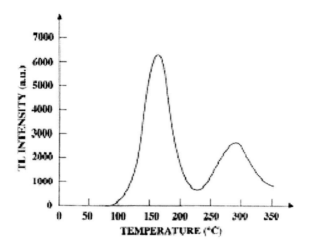

Fig.11. Glow curve of KMgF₃:La. H.R. at 10 Ks⁻¹.

- Lu-doped KMgF₃ [13]

This system was the last one studied. It was produced in three different dopant concentrations: 0.17, 0.34 and 0.66 mol%. The most intense TL was obtained wth the highest concentration: its glow curve is shown in **Fig.12** and its TL-dose response is given in **Fig.13**. The stability, over a period of 30 days, was very good.

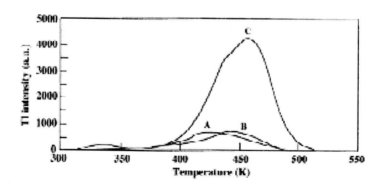

Fig.12. Glow curves of KMgF₃:Lu with different impurity concentrations. (A) 0.17 mol%, (B) 0.34 mol%, (C) 0.66 mol%.

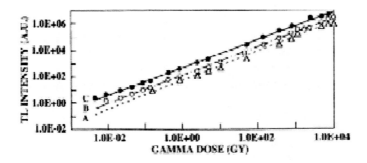

Fig.13. TL response of $KMgF_3$:Lu as a funtion of the various Lu concentrations.

Kinetics Parameters

The kinetics parameters of $KMgF_3$ have been studied for some of the previous preparations [12,13,14]. Table 1 lists the parameters obtained using various methods, i.e. Urbach (E_U) [15], Grossweiner (E_G) [16], Lushchik (E_L) [17], Halperin-Braner (E_{HB}) [18], Chen (E_C) [19] and the Computerized Glow Curve Deconvolution method (CGCD) [20].

Table 1. Kinetics parameters of perovskite

Peak N.o	Ce	b	E_U eV	$E_G/1^{st}$ eV	$E_G/2nd$ eV	$E_L/1st$ eV	$E_L/2^{nd}$ eV	$E_{HB}/1^{st}$ eV	$E_{HB}/2^{nd}$ eV	$E_{C\tau}$ eV	$E_{C\delta}$ eV	$E_{C\omega}$ eV	s s^{-1}	E eV CGCD
	0.24%	1.4	0.96	-	-	-	-	-	-	1.16	1.20	1.19	8×10^{11}	-
	0.5%	1.7	0.98	-	-	-	-	-	-	1.30	1.30	1.30	7×10^{12}	-
	1.0%	1.9	1.0	-	1.10	-	1.08	-	1.09	1.08	1.08	1.08	2×10^{10}	-
	1.5%	1.9	0.99	-	1.04	-	1.06	-	1.03	1.02	1.02	1.05	10^{10}	-

Peak N.o	Er	b	E_U eV	$E_G/1^{st}$ eV	$E_G/2nd$ eV	$E_L/1st$ eV	$E_L/2^{nd}$ eV	$E_{HB}/1^{st}$ eV	$E_{HB}/2^{nd}$ eV	$E_{C\tau}$ eV	$E_{C\delta}$ eV	$E_{C\omega}$ eV	s s^{-1}	E eV CGCD
	1.0%	1.9	1.09	-	1.51	-	1.50	-	1.53	1.52	1.50	1.52	3×10^{13}	-
	1.3%	1.0	1.08	1.03	-	1.01	-	1.04	-	1.05	1.09	1.07	10^9	-
	2%	1.8	0.98	-	1.16	-	1.20	-	1.16	1.14	1.15	1.15	2×10^{10}	-

Peak N.o	La	b	E_U eV	$E_G/1^{st}$ eV	$E_G/2nd$ eV	$E_L/1st$ eV	$E_L/2^{nd}$ eV	$E_{HB}/1^{st}$ eV	$E_{HB}/2^{nd}$ eV	$E_{C\tau}$ eV	$E_{C\delta}$ eV	$E_{C\omega}$ eV	s s^{-1}	E eV CGCD
1st peak	2%	2	-	-	1.04	-	1.03	-	1.04	1.04	1.03	1.04	6.9×10^{11}	-
2nd peak		1.9	-	-	1.21	-	1.26	-	1.20	1.18	1.21	1.20	3.4×10^{10}	-

Peak N.o	Lu	b	E_U eV	$E_G/1^{st}$ eV	$E_G/2nd$ eV	$E_L/1st$ eV	$E_L/2^{nd}$ eV	$E_{HB}/1^{st}$ eV	$E_{HB}/2^{nd}$ eV	$E_{C\tau}$ eV	$E_{C\delta}$ eV	$E_{C\omega}$ eV	s s^{-1}	E eV CGCD
1st peak		2	-	-	-	-	-	-	-	-	-	-	-	0.71
2nd peak		1	-	-	-	-	-	-	-	-	-	-	-	1.26
3rd peak	0.66%	2	-	-	-	-	-	-	-	-	-	-	-	1.81
4th peak		2	-	-	-	-	-	-	-	-	-	-	-	2.13
5th peak		2	-	-	-	-	-	-	-	-	-	-	-	2.33

Conclusions

The above reported data show that the perovskite-like KMgF$_3$, doped by various impurities, represents a real success in the search of efficient dosimetric materials.

The only drawback of this material is the presence of the natural isotope ^{40}K, which induces a TL signal superimposed to the TL signal due to the external irradiation. The net contribution of ^{40}K to the total dose has been estimated to be about 0.22 mGy/month [10,13]: this quantity is very little and in dosimetric applications as in clinical monitoring during radiological treatments does not create any serious problem.

References

[1]   Furetta C, Bacci C, Rispoli B, Sanipoli C and Scacco A. 1990. Luminescence and dosimetric performances of KMgF$_3$ crystals doped with metal impurity ions. *Rad. Prot. Dos.* **33**(1/4), 107

[2]   Bacci C, Furetta C, Ramogida G, Sanipoli C and Scacco A. 1993. Radiation dosimetry with Eu-doped crystals of KMF$_3$. *Physica. Medica IX (Suppl. 1),* **207**

[3]   Furetta C, Scacco A, Rabkin L M and Rudkovsky V N. 1993. *Stimolazione termica di luminescenza ed emissione elettronica in cristalli di KMgF$_3$ contenenti impurezze.* XXVIII Congresso Nazionale IRPA, Taormina, Italy.

[4]   Bacci C, Fioravanti S, Furetta C, Missori M, Ramogida G, Rossetti R, Sanipoli C and Scacco A. 1993. Photoluminescence and thermally stimulated luminescence in KMgF$_3$:Eu^{2+} crystals. *Rad. Prot. Dos.* **47**, 277

[5]   Scacco A, Furetta C, Bacci C, Ramogida G and Sanipoli C. 1994. Defects in γ-irradiated KMgF$_3$:Tl$^+$ crystals. *Nucl. Instr. Meth. Phys. Res.* **B91**, 223

[6]   Furetta C, Ramogida G, Scacco A, Martini M and Paravisi S. 1994. Spectroscopy of complex defects in crystals of KMgF$_3$:Tl$^+$. *J. Phys. Chem. Solids* **55**(11), 1337

[7]   Kitis G, Furetta C, Sanipoli C and Scacco A. 1996. Thermoluminescence properties of KMgF$_3$ doped with Pb, Cr and Ag. *Rad. Prot. Dos.* **65**(1/4), 93

[8]   Kitis G, Furetta C, Sanipoli C and Scacco A 1999. KMgF$_3$:Ce, an ultra-high sensitività thermoluminescent material. *Rad. Prot. Dos.* **82**(2), 151

[9]   Furetta C, Santopietro F, Sanipoli C and Kitis G. 2001. Thermoluminescent (TL) properties of the perovskite KMgF$_3$ activated by Ce and Er impurities. *Appl. Rad. Isot.* **55**, 857

[10] Le Masson N J M, Bos J J, Van Eijk C W E, Furetta C and Chaminade J P 2002. Optically and thermally stimulated luminescence of KMgF$_3$:Ce^{3+} and NaMgF$_3$:Ce^{3+}. *Rad. Prot. Dos.* **100**, 229

[11] Sepulveda F, Azorin J, Rivera T, Furetta C and Sanipoli C. *Thermoluminescence*

[12] Furetta C, Azorin J, Sepulveda F, Rivera T and Gonzalez P R. 2002. Thermoluminescence kinetic parameters of perovskite-like KMgF$_3$ activated by La ions. *J. Mat. Sc. Lett.* **21**, 1727

[13] Gonzalez PR, Furetta C, Azorin J, Rivera T, Kitis G, Sepulveda F and Sanipoli C. 2004. Thermoluminescence (TL) characterization of the perovskite-like KMgF$_3$ activated by Lu impurity. *J. Mat. Sc.* **39**, 1601

[14] Furetta C, Sanipoli C and Kitis G. 2001. Thermoluminescent kinetics of the perovskite KMgF$_3$ activated by Ce and Er impurities. *J. Phys. D: Appl. Phys.* **34**, 857

[15] Urbach F. 1948. Storage and release of light by phosphors. *Cornell Symposium*, **115**, Wiley, N.Y.

[16] Grossweiner L I. 1953. A note of the analysis of first order glow curves. *J. Appl. Phys.* **24**, 1306

[17] Lushchik C B. 1956. The investigation of trapping centres in crystals by the method of thermal bleaching. *Soviet Physics JEPT* **3**, 390

[18] Halperin A and Braner A A. 1960. Evaluation of thermal activation energies from glow curves. *Phys. Rev.* **117**, 408

[19] Chen R. 1969. Glow curves with general order kinetics. *J. Electrochem. Soc.* **116**, 1254

[20] Chen R and Kirsh Y. 1981. Analysis of thermally stimulated processes. *Pergamon Press*, GB

In: Trends in Materials Science Research
Editor: B.M. Caruta, pp. 157-190

ISBN: 1-59454-367-4
© 2006 Nova Science Publishers, Inc.

Chapter 7

THE EQUIVALENT SIMPLE CUBIC SYSTEM: A NEW TOOL TO MODEL REAL POWDER SYSTEMS UNDER COMPRESSION

J.M. Montes[1], F.G. Cuevas, J. Cintas,
J.A. Rodríguez and E.J. Herrera

Department of Mechanical and Materials Engineering, Escuela Superior de Ingenieros, Universidad de Sevilla, Camino de los Descubrimientos, s/n, 41092 Sevilla, Spain.

Abstract

In this work, a new way of approaching some aspects related to powder compaction is proposed. This new route consists on modeling a real powders system (particles of unequal size and form) by means of a system of spheres with simple cubic packing. This latter system suffers the same type of deformation that the real system (uniaxial, biaxial or triaxial), and possesses a porosity degree that in some aspects, makes it equivalent to the real one.

The evolution of particles shape, the effective contact area between particles and the effective path to be traveled into the powder aggregate by the electrical or thermal flow are of great interest in order to establish the connection between both systems (real and cubic). These topics are studied for the simple cubic system and the real system.

Initially, the deformation of the simple cubic system of spheres, subjected to uniaxial, biaxial and triaxial compression, is studied in detail. This study is essentially carried out by a geometric method (*the inflationary sphere model*), eluding the stress-strain problem. Equations to determine how the radius of the inflationary sphere increases as the porosity diminishes are deduced. Besides, equations determining the contact areas of a sphere with its neighbors, as a function of the porosity during the deformation, are established. These relationships suggest the introduction of a new concept: the *normalized* or *relative porosity* (the porosity to the initial simple-cubic-packing porosity ratio), that has been revealed extremely useful in subsequent studies.

A parallel study on real powder systems is carried out, focused on two questions: (i) the effective contact area determination (the real contact area that supports the truly applied pressure, or that serves as passage to the electric current, for example), and (ii) the effective path determination (the length of the shortest way, avoiding pores, that connects two points of

[1] E-mail address: jmontes@esi.us.es, Tel.: +34 954487305, Fax: +34 954460475

both bases of the compact, and that, in principle, would be the way to follow by the electrical or thermal flow). In both questions, use of the aforementioned relative porosity is made, now defined as the quotient between the porosity and the tap porosity of the starting powder. The most relevant proposed equations and predicted results are validated through experimental measurements on real powder aggregates and sintered powder compacts, as well as contrasted with results found in the literature.

After the study of the simple cubic and real systems, the equivalence equations to calculate the porosity that the simple cubic system should have to be considered equivalent to the real one, or vice versa, are established.

Finally, some problems, related to the compaction and electrical conduction of powder aggregates, are studied and solved by means of the new described technique. Proposed solutions to these problems are validated by experiences on real powder systems, and contrasted with data reported by other authors.

Introduction

Interest in the compaction of a powder aggregate spans several fields, as powder metallurgy, ceramic industry, pharmaceutical industry, geology, etc., since the external pressure repacks and deforms the particles into a higher density mass, and also imparts a new shape to the powder aggregate.

Numerous microscopic and macroscopic models have been developed to describe the deformation of a powder aggregate during compaction. Microscopic models [1-5] predict the response of the powder mass, to an external force, from the micromechanical behavior of its constituent particles. On the other hand, macroscopic models [6-12] approach the problem considering the powder aggregate as a continuous material. A good number of these models are limited to isostatic or quasi-isostatic loading conditions [9-12].

Although the microscopic models, a priori, can seem more realistic, they are not exempt of inconveniences. The main of them is their high computational cost, but also the necessary simplification of shapes and sizes distribution in the powder particles. Most models suppose monosize spherical particles following a random packing (Fig. 1a). More refined models suppose a distribution of spherical particles with different radii (Fig. 1b), and a few of them approach the problem supposing particles with ellipsoidal forms (Fig. 1c). Even this latter case is not a faithful reflection of reality. Powder particles have, in general, complex shapes (Fig. 1d). Therefore, given the impossibility of carrying out an accurate modeling of the particles form and distribution (prohibitive, also, in calculation time), it is not disheveled to appeal to macroscopic models.

From a theoretical point of view, in the study of a compaction process, the spherical shape of the particles is, of course, the most desirable. However, even in this case, the deformation study during compaction is extremely difficult. The compaction of spherical particles involves a transition in shape from spheres to polyhedral, as the pressure is increased. In the final stages of compaction, the material is usually considered a dense matrix with isolated pores, but modeling of the first stages (bellow 5% of porosity) requires an accurate knowledge of the evolution of the contact area and the distance between particle centers, as a function of the applied force.

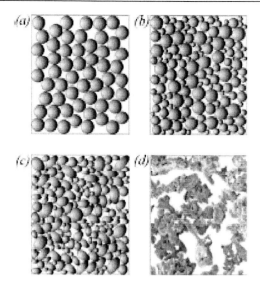

Figure 1. Different theoretical approximations to a powder system: (a) monosize spherical particles, (b) unequal size spherical particles, (c) unequal size ellipsoidal particles, and (d) unequal size and irregular forms particles in a real powder (reality is much more complex than models).

From all the possible packing arrangements, the simple cubic array is conceptually the simplest and, consequently, the easier to approach. Unfortunately, a system of spherical particles arranged in a simple cubic packing is very far from reality.

The objective of this work is to answer to the following question: can a real powders system under compression be represented by a spheres system with simple cubic packing? And, in affirmative case, which porosity should the orderly deforming spheres system have? The advantages of such a possibility are clear, since many problems, more easily soluble in the simple cubic system, would be solved there, and then the solution transferred to the real system. Fig. 2 shows this working method.

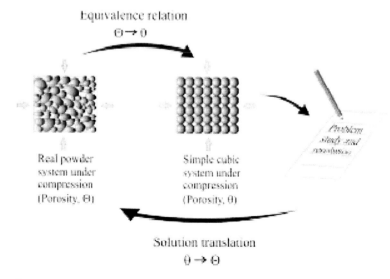

Figure 2. Work philosophy with the *Equivalent Simple Cubic System* (E.S.C.S.).

The method that will be developed in this chapter is not strictly macroscopic, neither microscopic; it can be described as a *hybrid technique*. Although the real powders are considered as a continuous system characterized by its porosity, the problem is solved in the *equivalent simple cubic system* (E.S.C.S.), where microscopic aspects are considered. Then, the solution is transferred to the real system.

In first place, the deformation of a simple cubic system of spheres will be studied in detail and, later on, some interesting aspects of the deformation of a real system of powders will be analyzed.

Simple Cubic System under Compression

Work Assumptions

Consider an ideal powder aggregate consisting of spherical particles of radius R_0 arranged in a simple cubic packing. The unit cell for such a system is depicted in Fig. 3(a), Fig. 4(a) and Fig. 5(a), where also the applied pressure is indicated in each case. Let us also assume the following:

(a)   *The constituent material of the spherical particles is ideally incompressible (i.e. its volume remains constant throughout the deformation) and shows a behavior absolutely plastic.*

(b)   *The unrestricted surface of the particle (the surface that is not in contact with any neighbor particle) is, at all times, a sphere [13, 14]. So, during compression, the initial sphere radius grows to such an extent that the volume enclosed by the circumscribing orthohedron continues to be identical to its initial volume (inflationary sphere model).*

(c)   *The acting compressive agent can supply as high a force as required to achieve the required deformation.*

Geometric Evolution During Uniaxial, Biaxial and Triaxial Compactions

The geometric evolution of the aforementioned unit cells, during the compaction process, can be described as follows: during uniaxial compression, the area of the base of the cube where the particle is inscribed, Fig. 3(a), remains constant ($4R_0^2$), while its height ($2l$) decreases in a gradual manner. For the biaxial case, it is the height of the cube where the particle is initially inscribed what remains constant ($2R_0$), Fig. 4(a), while the area of its base (initially, equals to $4R_0^2$, but in general, equals to $4l^2$) decreases. Finally, during triaxial compression, the area of a face of the cube ($4l^2$) where the particle is inscribed, Fig. 5(a), gradually decreases when the length of the cube face changes from $l = R_0$ to the final value of l. In all cases, compression ceases when l decreases to its minimum value, l_{final}, which ensures that the orthohedron volume equals that of the initial sphere ($\frac{4}{3}\pi R_0^3$).

Once the final value of l is known, the final radius of the inflationary spherical surface (that in this situation is totally external to the orthohedron, and only touches it in the eight vertices) can also be calculated, on the bases of simple geometric considerations.

On the other hand, during the deformation process the porosity of the unit cell is defined as

$$\theta = \frac{V_{void}}{V_{orthoedron}} = \frac{V_{orthoedron} - V_{occupied}}{V_{orthoedron}} = 1 - \frac{V_{occupied}}{V_{orthoedron}}$$

(1)

In addition, since the volume of the material ($V_{occupied}$) is assumed to remain constant ($\frac{4}{3}\pi R_0^3$), the porosity in the uniaxial, biaxial and triaxial cases can be calculated.

Expressions for porosity can be resumed in a single one as

$$\theta = 1 - \frac{\pi}{6}(R_0/l)^n$$

(2)

with $n = 1, 2$ and 3, for uniaxial, biaxial or triaxial compression, respectively.

Values for l_{final}, R_{final} and θ are compiled for the uniaxial, biaxial and triaxial cases in Table 1.

Table 1. Values of l_{final}, R_{final} and θ for uniaxial, biaxial and triaxial compaction processes

	Uniaxial	*Biaxial*	*Triaxial*
l_{final}	$\frac{\pi}{6}R_0$	$\left(\frac{\pi}{6}\right)^{\frac{1}{2}}R_0$	$\left(\frac{\pi}{6}\right)^{\frac{1}{3}}R_0$
R_{final}	$\sqrt{l_{final}^2 + 2R_0^2}$	$\sqrt{2l_{final}^2 + R_0^2}$	$\sqrt{3}\,l_{final}$
θ	$1 - \frac{\pi}{6}(R_0/l)$	$1 - \frac{\pi}{6}(R_0/l)^2$	$1 - \frac{\pi}{6}(R_0/l)^3$

The deformation of the initially spherical particle produces planar surfaces in the contacts with the six neighboring particles. Such surfaces are initially circular, Fig. 3(b), Fig. 4(b) and Fig. 5(b), however, the contact surfaces are not always circular. For the uniaxial case, the two contacts on the bases possess a larger radius than the four on the sides. Thus, when the diameter of the circle on the bases equals the length of the edge of the square base ($2R_0$), the basal and lateral circles touch each other [first critical situation, Fig. 3(c)]. A second critical situation arises when the lateral circles, truncated at the top and bottom, contact other lateral circles on the left and right, Fig. 3(e). For the biaxial case, the four contacts on the sides possess a larger radius than the two on the bases, Fig. 4(b). Thus, when the diameter of the circle on the sides equals the length of the edge of the square base ($2l$), the lateral circles touch each other [first critical situation, Fig. 4(c)]. A second critical situation arises when the lateral circles, truncated on the left and right, contact the basal circles, at the top and bottom, Fig. 4(e). Finally, for the triaxial case (where all the contacts are always identical), after the unique critical situation, Fig. 5(c), when the diameter of these circles exceeds the length of the square face edge ($2l$), the circles begin to truncate, Fig. 5(d).

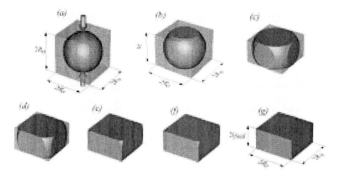

Figure 3. Different steps during uniaxial compression of an initially spherical particle. (*a*) Initial situation (contact zones are points). (*b*) Contact surfaces are circular. (*c*) First critical situation (for higher densification the circles on the bases are truncated in the X and Y directions, whereas those on the sides are truncated in the Z direction). (*d*) Intermediate situation. (*e*) Second critical situation (for higher densification lateral circles are truncated in vertical and horizontals directions). (*f*) Intermediate situation. *(g)* Final situation (the initial sphere has become an orthohedron).

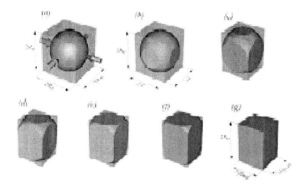

Figure 4. Different steps during biaxial compression of an initially spherical particle. (*a*) Initial situation (contact zones are points). (*b*) Contact surfaces are circular. (*c*) First critical situation (for higher densification the circles on the sides are truncated in the X and Y directions. (*d*) Intermediate situation. (*e*) Second critical situation (for higher densification lateral circles are truncated in the vertical and horizontals directions). (*f*) Intermediate situation. *(g)* Final situation (the initial sphere has become an orthohedron).

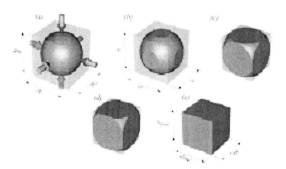

Figure 5. Different steps during triaxial compression of an initially spherical particle. (*a*) Initial situation (contact zones are points). (*b*) Contact surfaces are circular. (*c*) Critical situation. (*d*) Intermediate situation (the circles are truncated). (*e*) Final situation (the initial sphere has become a cube).

Numerical Strategy for the Calculation of Intermediate Situations

Different calculations are needed in order to know the exact geometric characteristics of any intermediate situations between the initial and final ones described above.

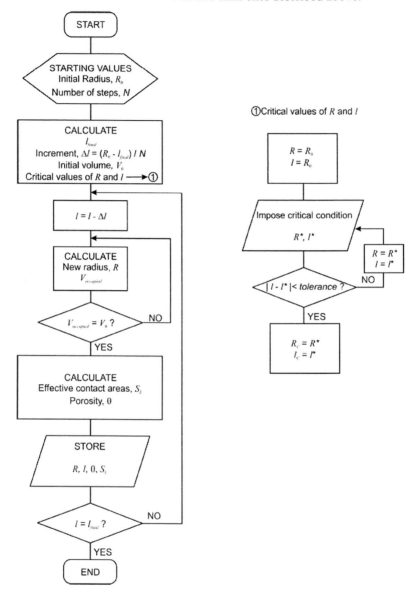

Figure 6. Flow diagram showing the numerical strategy for uniaxial, biaxial and triaxial compression problems resolution.

For a certain number of values of l, from $l = R_0$ to $l = l_{final}$, the radius of the inflationary sphere and the effective contact area can be calculated. These calculations have been carried out using a simple program developed in BASIC programming language. Fig. 6 shows the flow diagram describing the numerical strategy adopted.

In this resolution, calculations take into account the different critical situations shown in figures 3, 4 and 5. Critical values of the radius of the inflationary sphere, as well as dimensions of the orthohedron, are initially calculated based on geometric considerations by a successive approximations method, such that the volume occupied inside the orthohedron, $V_{occupied}$, equals the initial volume, V_0. Then, for each new value of l between the critical situations, the new radius of the inflationary sphere, R, is calculated in the same way. Known the radius of the inflationary sphere, the effective contact area, S_E, can be calculated (using the circular sector area equation when needed).

Once the effective contact area is known, the area ratio (S_E / S_N) can be calculated considering that $S_N = 4R_0^2$ in uniaxial compression, $S_N = 4lR_0$ in the biaxial case, and $S_N = 4l^2$ in the triaxial case. Then, the inflationary sphere radius (R) and the area ratio (S_E / S_N) can be related to the porosity, θ, in each situation from $l = R_0$ to $l = l_{final}$ (see values in Table 1).

The way to calculate the volume of the inflationary sphere inside the orthohedron as a function of l and R ($V_{occupied}$, in Fig. 6), needed to estimate by successive approximations the different critical values and R during the numerical resolution of the problem, is shown in Table 2. Before the first critical situation, this volume can be calculated using the formula of the volume of a spherical cap. After this, the volume remaining outside the orthohedron is difficult to calculate, because the spherical caps are truncated. This is even more complicated for the uniaxial and biaxial cases after the second critical situation, when the spherical caps are doubly truncated. Such formulae, which can not be found in the literature, are compiled in Table 3. All these different situations are implicitly considered in the numerical strategy of the flow diagram shown in Fig. 6.

Table 2. Volume calculation of the inflationary sphere inside the orthohedron ($V_{occupied}$), for uniaxial, biaxial and triaxial cases.

	Uniaxial	*Biaxial*	*Triaxial*
Initial situation Fig. 3(a), 4(a), 5(a)	$4/3\,\pi\,R_0^3$	$4/3\,\pi\,R_0^3$	$4/3\,\pi\,R_0^3$
Before 1st critical situation Fig. 3(b), 4(b), 5(b)	$V_{IS} - 2 \cdot V_{BSC} - 4 \cdot V_{LSC}$	$V_{IS} - 2 \cdot V_{BSC} - 4 \cdot V_{LSC}$	$V_{IS} - 6 \cdot V_{SC}$
After 1st critical situation Fig. 3(d), 4(d), 5(d)	$V_{IS} - 2 \cdot V_{BSC} - 4 \cdot V_{TLSC}$	$V_{IS} - 2 \cdot V_{BSC} - 2 \cdot V_{LSC} - 2 \cdot V_{TLSC}$	$V_{IS} - 2 \cdot V_{BSC} - 2 \cdot V_{TLSC} - 2 \cdot V_{DTLSC}$
After 2nd critical situation Fig. 3(f), 4(f)	$V_{IS} - 2 \cdot V_{BSC} - 2 \cdot V_{TLSC} - 2 \cdot V_{DTLSC}$	$V_{IS} - 2 \cdot V_{BSC} - 2 \cdot V_{TLSC} - 2 \cdot V_{DTLSC}$	-
Final situation Fig. 3(g), 4(g), 5(e)	$(2R_0)^2 \cdot 2l_{final}$	$2R_0 \cdot (2l_{final})^2$	$(2l_{final})^3$

IS, inflationary sphere; SC, spherical cup; BSC, basal spherical cap; LSC, lateral spherical cap; TLSC, truncated lateral spherical cap; and DTLSC, doubly truncated lateral spherical cap.

Table 3. Expressions to calculate spherical and truncated spherical caps volumes. Figures show the caps viewed from the top (circles), and the projection of the volume to be calculated (light grey).

Enclosed volume by planes $x = \pm k$, $y = \pm m$, $z = l$, and a sphere of radius R centered in the origin

Spherical cap

$$V = \frac{\pi}{3}(R-l)^2(2R+l)$$

Truncated spherical cap

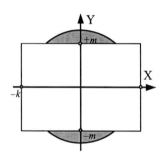

$$\beta = \sqrt{R^2 - l^2 - m^2}$$

$$V = \frac{2}{3}m(3R^2 - m^2)\tan^{-1}\left(\frac{\beta}{l}\right) + \frac{4}{3}R^3 \tan^{-1}\left(\frac{lm}{R\beta}\right) +$$

$$+ \frac{2}{3}l(l^2 - 3R^2)\tan^{-1}\left(\frac{l}{\beta}\right) - \frac{4}{3}lm\beta$$

Doubly truncated spherical cap

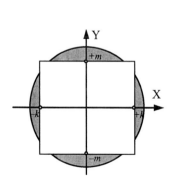

$$\beta = \sqrt{R^2 - l^2 - m^2} \quad \gamma = \sqrt{R^2 - l^2 - k^2}$$

$$V = \frac{2}{3}m(3R^2 - m^2)\tan^{-1}\left(\frac{\beta}{l}\right) + \frac{4}{3}R^3 \tan^{-1}\left(\frac{lm}{R\beta}\right) +$$

$$+ \frac{2}{3}l(l^2 - 3R^2)\tan^{-1}\left(\frac{l}{\beta}\right) - \frac{4}{3}lm\beta +$$

$$+ \frac{2}{3}k(3R^2 - k^2)\tan^{-1}\left(\frac{\gamma}{l}\right) + \frac{4}{3}R^3 \tan^{-1}\left(\frac{lk}{R\gamma}\right) +$$

$$+ \frac{2}{3}l(l^2 - 3R^2)\tan^{-1}\left(\frac{l}{\gamma}\right) - \frac{4}{3}lk\gamma - \frac{\pi}{3}(R-l)^2(2R+$$

Results: Inflationary Sphere Radius

Table 4 shows the values of the more salient parameters for uniaxial, biaxial and triaxial compression processes, namely, the initial, critical and final values of l, R and θ. A unity value was arbitrarily assigned to R_0 for these calculations.

The variation of the inflationary sphere radius (R) as a function of the *relative porosity* is shown in Fig. 7(a), (b) and (c). The relative porosity is defined as $\theta_R = \theta/\theta_M$, where θ_M is the maximum porosity of simple cubic packing (equals to its initial porosity, $1-\pi/6$).

Table 4. Half-height (l), radius (R) and porosity (θ) at initial, critical and final situations during uniaxial and triaxial compression.

UNIAXIAL	l	R	θ
Initial	1.0000	1.0000	0.4764
Critical 1	0.5878	1.1600	0.1092
Critical 2	0.5247	1.4142	0.0021
Final	0.5236	1.5080	0.0000
BIAXIAL	**l**	**R**	**θ**
Initial	1.0000	1.0000	0.4764
Critical 1	0.7761	1.0976	0.1307
Critical 2	0.7298	1.2380	0.0169
Final	0.7236	1.4308	0.0000
TRIAXIAL	**l**	**R**	**θ**
Initial	1.0000	1.0000	0.4764
Critical	0.8156	1.1534	0.0349
Final	0.8060	1.3960	0.0000

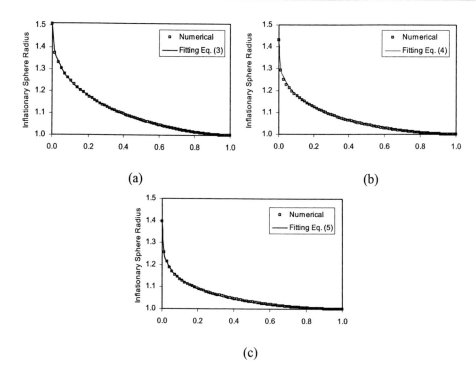

Figure 7. Inflationary sphere radius vs. relative porosity for *(a)* uniaxial and *(b)* biaxial and *(c)* triaxial compression, including numerical calculations, and fitting Eq. (3), Eq. (4) and Eq. (5), respectively.

It is possible to adjust the curves obtained from simulations by means of relatively simple analytical expressions. For example, for the uniaxial case, the inflationary sphere radius can be calculated as

$$R = R_0 \left[1 + (R_{final}/R_0 - 1)\left(1 - (\theta_R)^{0.40}\right)^{1.42} \right] \tag{3}$$

for the biaxial case,

$$R = R_0 \left[1 + (R_{final}/R_0 - 1)\left(1 - (\theta_R)^{0.35}\right)^{1.47} \right] \tag{4}$$

and for the triaxial case, with correlation coefficients higher than 0.99, in all cases.

Equations (3), (4) and (5) can be written jointly. If n represents the axial grade (1, for the uniaxial, 2 for the biaxial, and 3 for the triaxial case), then,

$$R = R_0 \left[1 + (R_{final}/R_0 - 1)\left(1 - (\theta_R)^{a}\right)^{b} \right] \tag{6}$$

with $a = 0.45 - 0.05n$ and $b = 1.37 + 0.05n$.

Results: Contact Areas Ratio

The variation of the area ratio (S_E / S_N) as a function of the relative porosity ($\theta_R = \theta/\theta_M$) is shown in Fig. 8.

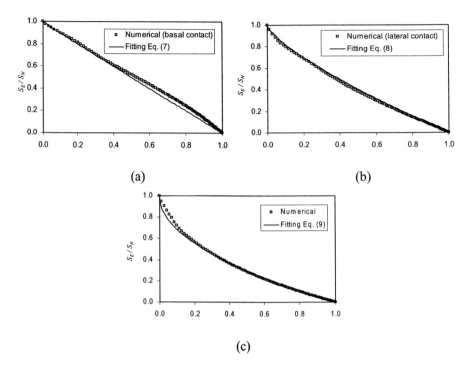

(a)   (b)

(c)

Figure 8. Area ratio vs. relative porosity for the big contacts in *(a)* uniaxial *(b)* biaxial and *(c)* triaxial compression, including numerical calculations, and fitting Eq. (7), Eq. (8) and Eq. (9), respectively.

The following simple equations relating porosity and effective area for the uniaxial, Eq. (7), biaxial Eq.(8) and triaxial case, Eq.(9) are proposed. These analytical expression describe the numerical curves in Fig. 8

$$S_E/S_N = 1 - \theta_R \tag{7}$$

$$S_E/S_N = 1 - \left(\theta_R\right)^{3/4} \tag{8}$$

$$S_E/S_N = 1 - \left(\theta_R\right)^{1/2} \tag{9}$$

As can be seen, the goodness of the fitting by Eq.(7), (8) and (9) to the calculated values (Numerical curves in Fig. 8) is reasonable (correlation coefficients of 0.989, 0.996 and 0.995, respectively). However, a slight difference is observed: fitting is better for lower relative porosities in uniaxial compression, and for higher θ_R in triaxial compression.

Considering equations (7), (8) and (9), a general expression relating S_E/S_N to θ_R can be proposed

$$S_E/S_N = 1 - (\theta_R)^{\frac{5-n}{4}}$$ (10)

with $n = 1$, 2 and 3, for uniaxial, biaxial and triaxial compression, respectively.

A similar study can be carried out with the small contact areas of the particle. Fig. 9 shows that small contact areas ratio (S_e/S_n) can also be related to the relative porosity (θ_R)

$$S_e/S_n = \left(1 - (\theta_R)^{\frac{5-n}{4}}\right)^{2.75}$$ (11)

with $n = 1$ for the uniaxial and $n = 2$ for the biaxial case. The correlation coefficients for this equation are 0.989 and 0.996 for the uniaxial and biaxial cases.

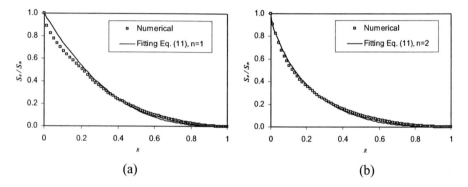

(a)   (b)

Figure 9. Small contact areas ratio vs. relative porosity for: (a) uniaxial and (b) biaxial compression, including numerical calculations and fitting Eq. (11).

Once the evolution of the theoretical simple cubic system is known, real powder systems are studied. Later on, both systems will be correlated.

Real Powder System under Compression

Effective Area

In practice, powder systems are much more complex than the simple cubic system. For this reason particle shape is not studied in this system, as particles are different from one to any other.

Both systems, real and simple cubic, share the fact that the initial porosity is different from 1. In real powders, the *minimum initial porosity of the loose (uncompressed) powder* is Θ_M. This is the porosity value corresponding to a perfect packing and is roughly equivalent to the *tap porosity*. The parameter Θ_M contains complex information about particles sizes, distribution and morphology of the powder; so, it somehow describes the pore structure of the powder system. Accordingly, Eq. (10) could be applied to real powder systems simply by substituting Θ_M for θ_M.

Thus, for the real powder systems, Eq. (10) can be written as

$$S_E/S_N = 1 - \left(\Theta_R\right)^{\frac{5-n}{4}} \qquad (12)$$

where $\Theta_R = \Theta/\Theta_M$, Θ is the porosity of the real system, and $n = 1$, 2 and 3, corresponding to uniaxial, biaxial and triaxial case, respectively. In Eq. (12), $S_E \rightarrow S_N$ when $\Theta_R \rightarrow 0$ (a fully dense sample), and $S_E \rightarrow 0$ when $\Theta_R \rightarrow 1$, i.e., when $\Theta \rightarrow \Theta_M$ (the upper limit of porosity) and the particles contacts are points.

A large number of expressions have been reported for this same calculation, some of which [15-18] are compiled in Table 5. All these expressions are a function of porosity and, also, all of them fulfill the condition that makes the effective area to increase when increasing densification. Besides, the effective area equals the nominal area when the porosity (Θ) becomes zero (*i.e.* $S_E \rightarrow S_N$ when $\Theta \rightarrow 0$). However, only the Helle equation [18] provides an accurate description of the starting situation of a powder aggregate, where the effective area should be virtually zero, *i.e.* $S_E \rightarrow 0$ when $\Theta \rightarrow \Theta_M$ (the initial porosity), because the initial contacts are points. In the remaining equations, except Spriggs one [17], $S_E \rightarrow 0$ when $\Theta \rightarrow 1$, a physically absurd limit for powder systems. This restricts the scope of these equations and excludes their use for the comprehensive description of compaction processes.

Table 5. Selected area-ratio expressions as a function of porosity, reported by different authors.

Author	Area ratio (S_E/S_N)
Early [15]	$1 - \Theta$
McClelland [16]	$1 - \Theta^{\frac{2}{3}}$
Spriggs [17]	$1/(1 + 2\Theta)$
Helle [18]	$(1-\Theta)^2(\Theta_0 - \Theta)/\Theta_0$

It is possible to validate Eq. (12) by means of experimental measurements of the effective contact area on a powder aggregate at different stages of a compression process. Spherical bronze powders (89/11 AK), from Ecka Granules (Fürth, Germany), with particle size between 45 and 63 μm and tap porosity of 0.4, have been cold uniaxial pressed at different porosity degrees. Micrographs of the specimens fracture surface are obtained by SE-SEM microscopy, and a very detailed selection of the contact areas is carried out. From these pictures, it can be measured the projection, on a normal plane to the applied load, of the contact areas, which is the effective area, S_E (Fig. 10). The process is finished by calculating the area percentage that the contacts represent, on a binary image where the contacts are shown. Fig. 11 shows a sequence of this process for one of the tested pressures.

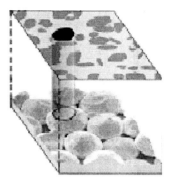

Figure 10. Determination of the effective area as a projection of the interparticle contacts on a normal plane to the compression force direction.

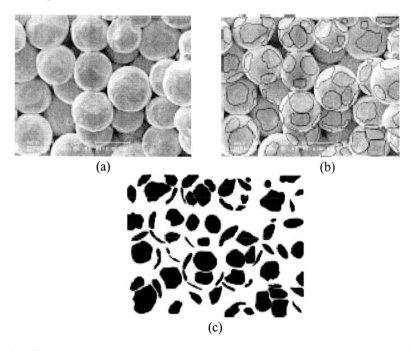

Figure 11. Different stages on the measurement of the effective area on powders pressed to the minimal tested pressure (400 MPa): (a) initial image, (b) selected zones of contact between particles, and (c) final binary image where the area ratio is automatically measured.

Fig. 12 represents the obtained results, and also includes similar results from other authors obtained on different powders [19, 20]. As can be seen, results are in very good agreement (correlation coefficient, $R^2 = 0.91$) with the prediction obtained from Eq. (12) for n = 1. Data from other authors are also in reasonably good agreement with the obtained results and the proposed equation.

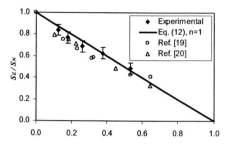

Figure 12. Experimental results of the effective area for different compacting pressures applied to spherical bronze powders. Data from Ref. [19] are represented assuming a tap porosity of 0.4 for the employed Ti-6Al-4V spherical powders. Data from Ref. [20] are represented assuming a tap porosity of 0.37 for the used powder, in experiences of closed die compression.

Effective Path

This variable, that was not studied in simple cubic systems as its knowledge is immediate, is of great interest in real systems. In an irregular pore structure the distance a fluid travels when passing through the pores exceeds the compact length in the flow direction. The *tortuosity* (ξ) is a measure of the ratio of actual flow distance (*effective path*, L_E) to the geometric compact length (L_N). This is an important concept in areas as vascular medicine, neurobiology, and stones and soils permeability, among others.

In this context, the variation of tortuosity ($\xi = L_E/L_N$) with porosity has been estimated in several studies [21-23]. Recent analyses suggest a relation of the type

$$\xi = A/\Theta^m \tag{13}$$

where A and m are constants. Table 6 gathers the values found for both parameters by different authors.

Table 6. Different values for A and m parameters in the tortuosity expression, Eq.(13), reported by some authors.

Author	A	m
Smith [21]	-	1
Meyer [22]	1.25	1.1
Smith [23]	0.9	1.3

Tortuosity is also an interesting quantity in powder metallurgy. When the flow takes place through the matrix instead of the pores (as it happens in electrical and thermal conduction of porous solids), a similar expression to the abovementioned can be used, substituting $1 - \Theta$ for Θ. So, for thermal and electrical flow, Eq. (13) is modified to:

$$\xi = A / (1 - \Theta)^m \qquad (14)$$

ξ now making reference to the ratio of effective path to the compact length, through the matrix.

However, Eq.(14) is not a good equation, because it does not satisfy the accurate boundaries conditions. In Eq.(14) , when $\Theta \rightarrow 0$, $\xi \rightarrow A$, instead of $\xi \rightarrow 1$, because the compact is fully dense. On the other hand, $\xi \rightarrow \infty$ when $\Theta \rightarrow 1$, however, for a powder compact, an infinite tortuosity should be reached when $\Theta \rightarrow \Theta_M$. In consequence, it would be interesting to have an equation satisfying the correct boundaries conditions. This equation can be obtained considering the arguments exposed in the following paragraphs.

Apart from size determination, volume fraction analysis is the oldest and the most commonly used stereological technique. The volume fraction of a phase can be determined form the data obtained by areal or linear analysis. Within the limits of the statistical scatter, in a porous specimen, the volume fraction of solid phase (i.e., the relative density, D) is equivalent to the areal and linear fractions [24]. The relative density equals $1 - \Theta$, where Θ is the porosity. Thus,

$$D = 1 - \Theta = \frac{V_0}{V_N} = \frac{S_0}{S_N} = \frac{L_0}{L_N} \qquad (15)$$

with S_0 the mean value of solid area, L_0 the mean value of solid segment, S_N the nominal area of a cross-section, and L_N the length of measuring in a porous sample (Fig. 13).

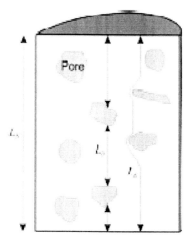

Figure 13. Representation of L_0, L_N and L_E in a cross-sectional surface of a porous material. L_N represents the geometrical length of the sample in the direction of the flow, L_0 is the solid fraction of L_N, and L_E is the actual path drawn by the flow.

From Eq.(15) it can be written

$$S_0 = S_N (1 - \Theta)$$ (16)

Comparing Eq.(16) with Eq.(12), it can be deduced that to calculate the effective area S_E (the transfer section in a real powder system, different to a cross section of the same system, S_0) the factor $(1 - \Theta)$ must be substituted by $\left(1 - (\Theta_R)^s\right)$, where $s = (5 - n)/4$.

In a similar way, from Eq.(15), it is also obtained that

$$L_N = \frac{L_0}{1 - \Theta}$$ (17)

However, as can be seen in Fig. 13, the true path length (or mean effective path, L_E) is bigger than and is related to L_N, but it is not directly related to L_0, that is smaller as does not consider the length of the pores. By an analogous reasoning, in order to calculate the true path length it is postulated that the factor $(1 - \Theta)$ must be substituted by $\left(1 - (\Theta_R)^s\right)$. Thus,

$$L_E = \frac{L_0}{1 - (\Theta_R)^s}$$ (18)

and as, according to Eq.(15), $L_0 = L_N (1 - \Theta)$, then, Eq.(18) is converted to

$$\boxed{L_E / L_N = \frac{1 - \Theta}{1 - (\Theta_R)^s}}$$ (19)

which expresses how the tortuosity (L_E / L_N) depends on the compact porosity and the tap porosity of the starting powder, through Θ_R. In Eq.(19), tortuosity tends to 1 when $\Theta_R \to 0$ (a fully dense sample), and it tends to infinite when $\Theta_R \to 1$, i.e., when $\Theta \to \Theta_M$ (the upper limit of porosity). The infinite value of tortuosity means an infinite value of L_E, what should be interpreted as the non-existence of viable paths connecting the considered points.

It is possible to validate Eq.(19) by measuring the shortest path between to points of a sintered compact. Such calculus can be carried out by means of the denominated *Pathfinding A* Algorithm* [25] with a post-processing that smoothes the drawn trajectory, maximizing its curvature radius. In order to make this, a BASIC language program has been developed. This program uses a binary image of a section of a sintered compact, drawing a big number of paths, avoiding pores, connecting different two extreme points, and finally, calculates the ratio of the mean length of those paths to the nominal length of the image. This final result is then related to the sample porosity, which can be calculated in the same image.

In order to validate Eq.(19), powder compacts of spherical Bronze and filament-shape Fe have been obtained by cold uniaxial pressing and furnace low-temperature sintering. This thermal treatment is necessary to maintain the compacts integrity during their preparation and study, but pores structure is not modified. Fig. 14 shows a sequence of the followed process for each compact.

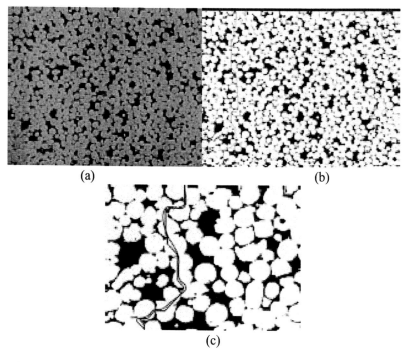

(a)   (b)

(c)

Figure 14. Different stages on the measurement of the effective path on a particular compact of bronze (with a porosity of 0.25): (a) initial image, (b) binarized image, and drawing of the shortest path via *Pathfinding A* Algorithm*, and (c) final smoothing of the trajectory (in general, this post-processing increases the trajectory length, but the result is more realistic).

Experimental results and the theoretical prediction obtained from Eq. (19) for $s = 1$ (uniaxial compression) are represented in Fig. 15.

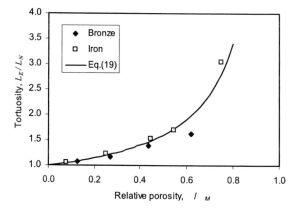

Figure 15. Experimental results of tortuosity for several sintered compacts (bronze and iron) with different porosity degrees and theoretical prediction according to Eq. (19).

As can be seen, fitting between theoretical and experimental results is satisfactory.

Equivalence Relation between Simple Cubic and Real Systems

In order to establish this relation, the crucial question is: what porosity (θ) must a simple cubic system consisting of deformed spheres have, to be equivalent to a real powder system with a porosity Θ and a tap porosity Θ_M? If both systems are composed of the same material and both have an identical geometrical aspect (equal S_N and equal L_N, or simply, identical ratio S_N/L_N) it can be supposed that this equivalence means the equality of the quotient between the effective area and the effective path length, i.e.,

$$\left[\frac{S_E}{L_E}\right]_{E.S.C.S.} = \left[\frac{S_E}{L_E}\right]_{Real} \tag{20}$$

It can be easily proved that this equivalence means that both systems have the same conductivity, and, in consequence, the study of problems related to thermal or electrical conductivities should be susceptible of being studied with this relation. However, problems where other physical variables are involved could or not be susceptible of being studied by this same hypothesis, and will be experimentally checked when the equivalent simple cubic system is employed.

Thus, taking into account Eq. (10), Eq. (12) and Eq. (19), and that $L_E = L_N$ for the simple cubic system, Eq.(20) can be written as:

$$\frac{S_N\left(1-(\theta_R)^s\right)}{L_N} = \frac{S_N\left(1-(\Theta_R)^s\right)}{L_N(1-\Theta)\left(1-(\Theta_R)^s\right)^{-1}} \tag{21}$$

with $s = (5-n)/4$, and then,

$$\left(1-(\theta_R)^s\right) = \frac{\left(1-(\Theta_R)^s\right)^2}{(1-\Theta)} \tag{22}$$

The right member of this equation can be approximated by a function of the type

$$\left(1-(\Theta_R)^s\right)^t \tag{23}$$

where the exponent t is a parameter depending on Θ_M. The advantage of this change is that only the variable Θ_R intervenes in the equation. Fig. 16 represents the parameter t as function of Θ_M for uniaxial, biaxial and triaxial cases. The points have been obtained by fitting Eq.(23) to the function of the right member of the Eq. (22), for different values of Θ_M, by the least squares method.

The t exponent can be expressed, with correlation coefficients better than 0.999, by a equation of the type

$$t = c_n + (2 - c_n)(1 - \Theta_M)^{0.8} \tag{24}$$

where c_n has a value of 1, 1.2536 and 1.5397 for $n = 1$, 2 and 3 (uniaxial, biaxial and triaxial cases), respectively.

Considering this, Eq.(22) is transformed to

$$\left(1 - (\theta_R)^s\right) = \left(1 - (\Theta_R)^s\right)^t \tag{25}$$

hence,

$$\theta_R = \left[1 - \left(1 - (\Theta_R)^s\right)^t\right]^{1/s} \tag{26}$$

with $\theta_R = \theta/\theta_M$, $\Theta_R = \Theta/\Theta_M$, and $s = (5 - n)/4$. Eq.(26) is the *equivalence relation*. This equation allows estimating the relative porosity of the equivalent simple cubic system (θ_R) known the relative porosity of the real system (Θ_R).

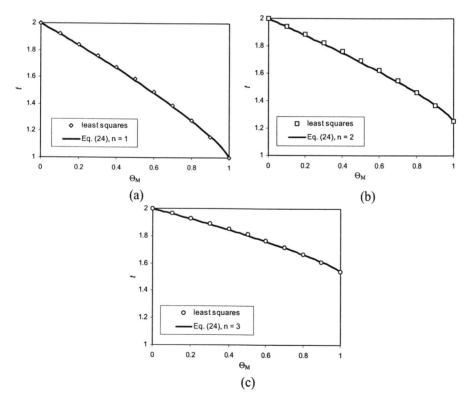

(a)

(b)

(c)

Figure 16. Exponent *t* as a function of Θ_M for (a) uniaxial, (b) biaxial and (c) triaxial compression. Spots shown in plots correspond to *t* values obtained by fitting Eq. (23) to the right member of the Eq. (22), for different values of Θ_M, by the least squares method. Solid lines are obtained from Eq. (24).

Examples: Applications of the Equivalent Simple Cubic System

Pressure-Density Law in Uniaxial Compression

Powder compaction has assumed an important role in the manufacture of many ceramic and metallurgical products. Many equations have been proposed to relate porosity (or relative density) to pressure [26-31]. Table 7 presents some of these equations, expressed in terms of the porosity, Θ.

The main critic to be done to equations in Table 7 is not a bad agreement to experimental data, as all of them have two fitting parameters, and the correlation coefficients are normally very good. However, these parameters do not have any clear physical meaning and, in consequence, can not be determined by direct measurements. Very few authors [20, 32-34] have attempted to elaborate new models based on the powder mechanical behavior. The objective now is to deduce, through the equivalent simple cubic system, an equation whose parameters have a clear physical meaning, and can be directly measured. The difficulty is big: can a powder compaction process be modeled from the macroscopic mechanical characteristics of the constituent material or, on the contrary, mechanical measurements on powder particles in the microscopic scale are necessary? An exhaustive study of this problem would be too extensive. Here, it will only be shown how by the E.S.C.S. can be easily obtained a compaction equation, with a reasonable agreement to experimental data and a more elaborated model by other author [20].

Table 7. Several compaction equations where P_N is the applied pressure, Θ is the porosity, Θ_0 is the porosity of the loose powder without pressure application, Θ_∞ is the minimum porosity to be reached, and c_1 to c_{12} are constants.

Author	Year	Equation
Balshin [26]	1938	$\ln(P_N) = -c_1/(1-\Theta) + c_2$
Heckel [27]	1961	$\ln(1/\Theta) = c_3 P_N + c_4$
Kawakita [28]	1970	$(1-\Theta)/(\Theta_0 - \Theta) = c_5/P_N + c_6$
Ge [29]	1995	$\log[\ln(1/\Theta)] = c_7 \log(P_N) + c_8$
Panelli [30]	1998	$\ln(1/\Theta) = c_9 P_N^{1/2} + c_{10}$
Secondi [31]	2002	$\ln((\Theta - \Theta_\infty)/(\Theta_0 - \Theta_\infty)) = -c_{11} P_N^{c_{12}}$

In first place, a relation between the applied pressure (P_N) and the porosity (θ) is going to be deduced for the simple cubic system. This is easily done as its geometry is known during the deformation process. Afterwards, the behavior of a real system will be deduced through the equivalence relation between both systems and the aforementioned equation.

The stress on a unit cell during a compaction process can be calculated as the quotient between the applied load (F) and the effective area where this load is applied (S_E). Thus, considering Eq. (7),

$$\sigma = \frac{F}{S_E} = \frac{F}{S_N(1-\theta_R)} = P_N(1-\theta_R)^{-1} \qquad (27)$$

On the other hand, the plastic behavior of the material can be described by [20]:

$$\sigma = \sigma_y(1+h\varepsilon) \qquad (28)$$

where σ_y and h are material characteristic parameters (σ_y is the yield stress and h is the *linear strain hardening coefficient*).

In addition, the true deformation of the simple cubic unit cell, subjected to uniaxial compression, where the initial porosity is $\theta_M = 1 - \pi/6$, can be defined as:

$$\varepsilon = \ln\left(\frac{2R_0}{2l}\right) = \ln\left(\frac{R_0}{l}\right) \qquad (29)$$

where $2l$ is the unit cell height.

According to Eq. (2), with $n = 1$,

$$l = \tfrac{\pi}{6}\frac{R_0}{(1-\theta)} \qquad (30)$$

thus, Eq.(29) can be written as

$$\varepsilon = \ln\left(\tfrac{6}{\pi}(1-\theta)\right) \qquad (31)$$

Introducing Eq. (27) and (31) in Eq. (28), it is obtained

$$P_N(1-\theta_R)^{-1} = \sigma_y\left[1 + h\ln\left(\tfrac{6}{\pi}(1-\theta)\right)\right] \qquad (32)'$$

and then,

$$P_N = \sigma_y\left[1 + h\ln\left(\tfrac{6}{\pi}(1-\theta)\right)\right](1-\theta_R) \qquad (33)$$

This is the equation relating P_N and θ for the simple cubic system. In order to obtain corresponding relation for the real system, the equivalence relation in Eq. (26) has to be used. For the uniaxial case this equivalence relation is reduced to:

$$\theta_R = 1 - (1-\Theta_R)^t \qquad (34)$$

and taking into account that $\theta_M = 1 - \pi/6$ and $\theta = \theta_R \theta_M$, the following expression is obtained for the real system, expressing the relation between pressure, P_N, and porosity, Θ, for powders under uniaxial compression.

$$P_N = \sigma_y \left[1 + h \ln\left(1 + (\tfrac{6}{\pi} - 1)(1 - \Theta/\Theta_M)^t\right)\right](1 - \Theta/\Theta_M)^t \qquad (35)$$

where, according to Eq. (24) with $n = 1$, $t = 1 + (1 - \Theta_M)^{0.8}$.

In Eq.(35), $P_N \to P_{NM\acute{a}x} = \sigma_y \left[1 + p \ln(6/\pi)\right]$ when $\Theta \to 0$. When $\Theta \to \Theta_M$, $P_N \to 0$, as can be expected.

The obtained equation is mathematically some more complex than those shown in Table 7, on the other hand, an important advantage should be considered: all parameters in the equation have a physical meaning. Some of them represent characteristics of the powder morphology and grain size distribution (Θ_M and t), others, powder material mechanics characteristics (σ_y and h) that, in principle, can be macroscopically measured.

Fig. 17 shows the *nominal pressure* vs. *compact porosity* curve obtained from Eq. (35), the theoretical prediction, and the experimental results reported by [20]. Experimental data correspond to a Cu powder with a tap porosity of 0.36 and which stress-strain characteristic $\sigma(\varepsilon)$ was determined from incremental compression tests performed on forged, fully dense specimens. As can be seen, fitting between theoretical prediction of Eq. (35) and experimental data is acceptable, even some better than the theoretical prediction suggested by [20] with a more complex model (correlation coefficients R^2 of 0.79 y 0.65, respectively).

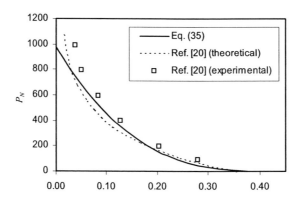

Figure 17. Experimental densification curves and theoretical predictions from Eq. (35) and the model proposed by Fischmeister *et al* [20] corresponding to a Cu powder with a tap porosity of 0.36.

It is then proved that the hypothesis assumed to obtain the equivalence relations between the simple cubic and real systems, it is also proper to model mechanical nature problems in powders, and not only electrical nature problems.

Electrical Resistivity of Powder Aggregates Consisting of Film-Covered Particles under Uniaxial Compression

The objective now is calculating the electrical resistivity of an actual powder system consisting of oxide-covered particles. We shall assume the powder particles to have an average radius R_0, δ to be the average thickness of their oxide films, and Θ_R the relative porosity of the system as a whole.

Consider a simple cubic system consisting of oxide-covered particles having a relative porosity θ_R, as calculated from the equivalence relation, Eq.(26). Let us assume the electrical resistance of the unit cell to be identical to that of a cylinder connecting the centers of two adjacent particles and having a cross-sectional area coincident with the contact area (Fig. 18).

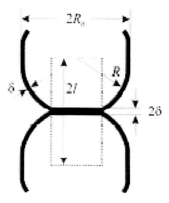

Figure 18. Detail of the contact area between two particles in the simple cubic system subjected to uniaxial compression. The value R corresponds to the radius of the inflationary sphere. The thickness of the dielectric layer is δ. The distance between two particle centers is $2l$. The electrical resistance of unit cell can be calculated as the resistance of cylinder drawn in dashed line.

If $2l$ is the distance between particle centers and δ the thickness of the dielectric layer, then the resistance of the cylinder, R_c, will be given by

$$R_c = \rho_M \frac{2(l-\delta)}{S_E} + \rho_X \frac{2\delta}{S_E} = \frac{2}{S_E}\left((l-\delta)\rho_M + \delta\rho_X\right) \tag{36}$$

where ρ_M is the resistivity of the metal and ρ_X that of the oxide film. The effective resistivity of the cell, ρ_E^*, and, by extension, that for the whole sample, will thus be

$$\rho_E^* = R_c \frac{S_N}{2l} = \frac{(l-\delta)\rho_M + \delta\rho_X}{l}\left(\frac{S_N}{S_E}\right) \tag{37}$$

therefore, based on Eq.(7),

$$\rho_E^* = \frac{(1-f)\rho_M + f\rho_X}{1-\theta_R} \tag{38}$$

where $f = \delta/l$. As l is most often much greater than δ ($f \sim 10^{-3}$), Eq. (38) can be simplified to

$$\rho_E^* = \frac{\rho_M + f\rho_X}{1-\theta_R} \tag{39}$$

Also, according to Eq.(2) with $n = 1$,

$$l = \tfrac{\pi}{6}\frac{R_0}{(1-\theta)} \tag{40}$$

hence, since $\theta = \theta_R \cdot \theta_M$ and $\theta_M = 1 - \tfrac{\pi}{6}$,

$$f = \frac{\delta}{l} = \tfrac{6}{\pi}\frac{\delta}{R_0}\left(1 - (1 - \tfrac{\pi}{6})\theta_R\right) \tag{41}$$

substituting Eq. (41) into Eq. (39) gives

$$\rho_E^* = \left(\rho_M + \tfrac{6}{\pi}\rho_X \frac{\delta}{R_0}\left(1 - (1 - \tfrac{\pi}{6})\theta_R\right)\right)\frac{1}{(1-\theta_R)} \tag{42}$$

Finally, the results for the cubic system are transposed to the real system. This can be done by substituting Eq. (26), with $s = 1$, into Eq. (42):

$$\boxed{\rho_E = \left[\rho_M + \beta\left(1 + (\tfrac{6}{\pi} - 1)(1 - \Theta/\Theta_M)^t\right)\right](1 - \Theta/\Theta_M)^{-t}} \tag{43}$$

where

$$\boxed{\beta = \rho_X \cdot \delta/R_0} \tag{44}$$

and, according to Eq. (24) with $n = 1$, $t = 1 + (1 - \Theta_M)^{0.8}$.

As the resistivity of the metal is most often much lower than that of the oxide film (*i.e.* $\rho_M \ll \rho_X$), the following simpler equation can be used instead of Eq. (43):

$$\rho_E \approx \beta \left(\left(1 - \Theta/\Theta_M \right)^{-t} + \frac{6}{\pi} - 1 \right)$$

(45)

In Eq. (45), ρ_E tends to $6\beta/\pi$ when $\Theta \rightarrow 0$, and tends to ∞ when $\Theta \rightarrow \Theta_M$. These limits make advisable to introduce a new dimensionless quantity, F, defined as

$$F \equiv \left(\rho_E/\beta - \frac{6}{\pi} + 1 \right)^{\gamma_t} = \left(1 - \Theta/\Theta_M\right)^{-1}$$

(46)

where F tends to 1 when $\Theta \rightarrow 0$ and tends to ∞ when $\Theta \rightarrow \Theta_M$. The advantage of plotting F against Θ_R $(=\Theta/\Theta_M)$ is that it allows the data for different materials to be displayed in a single graph.

Eq. (46) was validated through the measurement of the electrical resistance of columns of different powders. Different porosities were attained increasing the load on the powder column. Three different elemental metallic powders were employed (iron, nickel and titanium), and measurements were carried out at room temperature (ca. 25 °C). The experimental data were fitted trough Eq. (46), by the least squares method, and the β values yielding the best fit were obtained. Fig. 19 shows the experimental and theoretical curves of F vs. relative porosity.

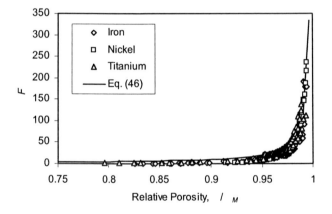

Figure 19. Experimental and theoretical variation of the dimensionless function F with the relative porosity of the different powders studied. Solid line corresponds to Eq. (46).

As can be seen, consistency was quite good in all cases (the correlation coefficient, R^2, was always greater than 0.77). The β values used, and the corresponding ρ_X values obtained from Eq. (44), are listed in Table 8 together with the nominal thicknesses of the oxide films, which were taken from the literature [35].

Table 8. Powders characteristic parameters —tap porosity (Θ_M), mean particle radius (R_0), mean thickness of the oxide film (δ)—, the fitting parameter β, and the predicted values of oxides resistivities, ρ_X.

Powder	Θ_M	R_0 (μm)	δ (Å)	β (Ω·m)	ρ_X (Ω·m)
Iron	0.56	39.21	35	$1.57 \cdot 10^{-5}$	$1.759 \cdot 10^{-1}$
Nickel	0.90	0.52	40	$5.04 \cdot 10^{-5}$	$6.552 \cdot 10^{-3}$
Titanium	0.71	72.59	55	$1.53 \cdot 10^{-5}$	$2.019 \cdot 10^{-1}$

The calculated ρ_X values shown in Table 8 cannot be compared with the tabulated values for different oxide species as determined from measurements of massive samples. Obviously, the thin films covering the oxide particles must behave in a rather different manner depending on their polycrystalline or amorphous nature, the presence of vacancies or doping impurities and a number of other factors that may substantially alter their resistivity. Therefore, the proposed Eq. (43) and Eq.(45) provide a potential means for the indirect estimation of the average resistivity of oxide films covering metal powder particles.

Thermal Power Generated During the First Instants in the Electrical Resistance Sintering.

The *electrical resistance sintering* technique (E.R.S.) consists in the passage of a high intensity electric current through a powder mass which is simultaneously subjected to compression. This technique was already described in 1933 by Taylor [36], although its systematic study was not carried out until some years later by Lenel [37], who denominated it *'Resistance Sintering under Pressure'*. Later on, it has been object of numerous studies, mainly by Japanese and old U.S.S.R. researchers [38-46]. However, in the actual moment, the theoretical treatment of this technique is not the adequate. Difficulties to attain this are remarkable, mainly, in the modeling of the first instants of the process [48].

In the form of E.R.S. treated in this chapter, the process starts applying a pressure on the powder column, with no electric current passing through it. The powder column acquires, after this, an initial porosity Θ_0. After this initial pressing stage, the current begins to flow and the powder start heating. It has been described [48, 49] that, for metallic powders, during the first instants the electric current is acting, the powder aggregate porosity does not almost decrease; however, the aggregate global electrical resistance decreases drastically. During these first instants, the huge intensity current through the system only find very narrow ways to pass (the interparticles contact areas), so that the metal adjacent to these contact areas can even melt [47] (Fig. 20). This liquid phase does not spread on the particles surfaces, since it is retained by the oxide layer covering the metallic powder particles. In a later stage, when the amount of liquid phase is big enough, the oxide layer fracture due to the pressure effect. Then, the liquid phase is liberated and spreads, acting as a lubricant, facilitating particles rearranging mechanisms, responsible of the powder densification.

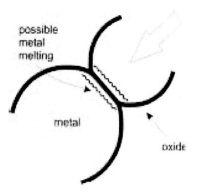

Figure 20. Detail of the contact zone between two powder particles passed through by a high intensity electric current. Under certain conditions, the metal close to the contact zone can melt.

The theoretical modeling of the first moments of the electrical sintering requires the heat generation and transmission knowledge in the inner of the powder mass, subjected to uniaxial compression. Even considering constant the powder column porosity, the problem is extraordinarily complex due to the shapes and sizes of the powder particles, the presence of an oxide layer around them, and the possible partial metal melting in the contacts between particles. It is, then, a complex problem where could be useful the use of the E.S.C.S. This section purpose is not to solve the whole heat conduction equation, but calculating the heat generation term due to the Joule effect; this is, the thermal power generated per volume unit. The equation total resolution, a complex problem, can be found in the literature [49].

When the powder column, subjected to a uniaxial load, has reached a porosity Θ_0, the cubic system porosity must be, considering the equivalent relation, Eq.(26), with $s = 1$:

$$\theta_0 = \theta_M \left[1 - \left(1 - \Theta_0 / \Theta_M \right)^t \right] \tag{47}$$

where, according to Eq. (24) with $n = 1$, $t = 1 + \left(1 - \Theta_M \right)^{0.8}$.

If an effective current intensity depending on time, $I(\tau)$, passes through the powder aggregate, with a nominal section S_N, then the current intensity passing through each basal contact between two particles of the E.S.C.S., $I_C(\tau)$, is:

$$I_C(\tau) = 4R_0^2 \left(I(\tau)/S_N \right) = 4R_0^2 \, j_N(\tau) \tag{48}$$

because the particles nominal section in the cubic system is $4R_0^2$. The term $j_N(\tau) = I(\tau)/S_N$ represents the *nominal current surface density*.

The electrical resistance of the dielectric layer in the contact area, with a thickness 2δ and an area $4R_0^2(1 - \theta_0/\theta_M)$, can be easily calculated, as:

$$R_X = \rho_X(T) \frac{2\delta}{4R_0^2 \left(1 - \theta_0/\theta_M \right)} \tag{49}$$

with ρ_X the oxide resistivity (a function of the temperature T). The thermal dependence of the electrical resistivity of an oxide is due to various factors (including the metal-dielectric contact) and it is not always well established. The charge transport mechanisms through thin-film isolators (with thicknesses bellow 1 micron), and phenomena associated to the dielectric breakdown, are complex and usually linked [47]. When the main process is a purely ohmic conduction, the resistivity shows a typical exponential thermal dependence, which is more complicated when the conduction mechanism is different. Anyway, it is going to be assumed that the function $\rho_X = \rho_X(T)$ is perfectly established.

Considering this, the thermal power generated in the contact by the oxide results

$$P_X(T,\tau) = I_C^2(\tau)R_X = \frac{8R_0^2\delta\rho_X(T)}{(1-\theta_0/\theta_M)}j_N^2(\tau) \tag{50}$$

It can be now supposed that this power is uniformly distributed on a semi-spherical layer of volume $V_X = 2\pi(R_0-\delta)^2\delta$. As the oxide layer thickness (a few nanometers) is much lower than the mean particles radius, this is, $\delta << R_0$, and then, $V_X \approx 2\pi R_0^2\delta$. So, the generated thermal power per volume unit, J_X, can be expressed as:

$$J_X(T,\tau) = \frac{P_X}{V_X} = \frac{4\rho_X(T)}{\pi(1-\theta_0/\theta_M)}j_N^2(\tau) \tag{51}$$

On the other hand, not only the power generated in the oxide layer, but also the metal contributes, in a lower amount, to the system heating. For simplicity, it is going to be considered that only the metal into the cylinder containing the basal contacts generates heat, however this heat is distributed in the whole particle.

A differential thickness disc of this cylinder should have an electrical resistance given by,

$$dR_M = \frac{\rho_M(T)dz}{4R_0^2(1-\theta_0/\theta_M)} \tag{52}$$

And then, the thermal power generated by this differential volume can be calculated as

$$dP_M = I_C^2(\tau)dR_M = \frac{4R_0^2\rho_M(T)dz}{(1-\theta_0/\theta_M)}j_N^2(\tau) \tag{53}$$

Considering that the heat generated by this disc is distributed by the whole particle section, that is, it is distributed in a differential volume

$$dV = S(z)dz \tag{54}$$

where the section $S(z)$, Fig. 21, can be a circle or a truncated circle, depending on the value of the coordinate z.

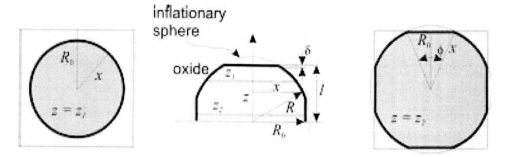

Figure 21. Different cross-sectional areas of a particle in the simple cubic system depending on the z value. For z = z1 (picture on the left), the particle section is fully circular. However, for z = z2 (on the right), the section is a circle truncated by the orthoedron in which the particle is inscribed.

Mathematically,

$$S(z) = \begin{cases} \pi x^2 & if \quad x \leq R_0 \\ x^2(\pi - 2(\phi - \sin\phi)) & if \quad x > R_0 \end{cases} \tag{55}$$

with $x = \sqrt{R^2 - z^2}$ and $\phi = 2\cos^{-1}(R_0/x)$. The radius R represents the radius of the inflationary sphere that can be known from Eq. (6) with $n = 1$, that is,

$$R = R_0 \left[1 + 0.508\left(1 - (\theta_0/\theta_M)^{0.40}\right)^{1.42} \right] \tag{56}$$

Then, the power generated in the metal per unit volume is:

$$J_M(z,T,\tau) = \frac{dP_M}{S(z)dz} = \frac{4R_0^2 \rho_M(T)}{S(z)(1-\theta_0/\theta_M)} j_N^2(\tau) \tag{57}$$

And the Joule effect term in the heat conduction equation can be expressed as:

$$J(z,T,\tau) = \begin{cases} \dfrac{4\rho_X(T)}{\pi(1-\theta_0/\theta_M)} j_N^2(\tau) & if \quad l-\delta \leq z < l \\[4mm] \dfrac{4R_0^2 \rho_M(T)}{S(z)(1-\theta_0/\theta_M)} j_N^2(\tau) & if \quad 0 \leq z < l-\delta \end{cases} \tag{58}$$

where l represents half-high of the unit cell (Fig. 21) and $S(z)$ is defined in Eq.(55).

Finally, in terms of the real system, according to Eq.(47), Eq. (58) can be written as

$$J(z,T,\tau) = \begin{cases} \dfrac{4\rho_X(T)}{\pi(1-\Theta_0/\Theta_M)^t}\, j_N^2(\tau) & if \;\; l-\delta \le z < l \\[4mm] \dfrac{4R_0^2\rho_M(T)}{S(z)(1-\Theta_0/\Theta_M)^t}\, j_N^2(\tau) & if \;\; 0 \le z < l-\delta \end{cases} \tag{59}$$

It is not possible to approach the direct validation of this expression. This section purpose has only been to illustrate, once more, how the E.S.C.S. allows studying very different problems related to powder compression.

Conclusions

A new way of approaching some aspects related to powder compaction is proposed in this work. This new method consists on modeling a real powder system under compression by means of a system of monosized deforming spheres with a simple cubic packing. This latter system is subjected to the same kind of deformation that the real system (uniaxial, biaxial or triaxial), and possesses a porosity degree which makes it equivalent to the real one. Thus, the real system, constituted by particles of unequal size and form, can be modeled by a simple cubic arrangement of deforming spheres. This working way has been called in this paper "Equivalent Simple Cubic System Technique". The most relevant hypothesis used in the theoretical approach are validated through experimental measurements on real powder aggregates and sintered compacts, as well as through computer simulations. Eventually, information reported in the literature is used.

Three problems related to the compaction and electrical conduction on powder aggregates are studied and solved to illustrate the power and applicability of this new technique. Proposed solutions for two of these problems are validated on real powder systems by experiences carried out by the authors or found in the literature.

Acknowledgements

The authors are grateful to FEDER/MCyT (Madrid) for funding this research within the framework of Project DPI2003–01213.

References

[1] Zavaliangos, A. *Materials Research Society Symposia Proceedings* 2002, 731, 169-176.
[2] Jagota, A. *Materials Research Society Symposiium Proceedings on Computational Methods in Materials Science* 1992, 278, 293-298.
[3] Dellis, C.; Bouvard, D.; Stutz, P. *Hot Isostatic Pressing' 93 Proceedings* 1994, 53-60.

[4]		Nair, S.V.; Tien, J.K. *Metall. Trans. A* 1987, 18, 97-107.

[5]		German, R.M., Particle packing characteristics; *Metal Powder Industries Federation*: Princeton, NJ, 1989.

[6]		Zavaliangos, A. *Int. J. Powder Metall.* 2002, 38(2), 27-39.

[7]		PM MODNET computer modelling group *Powder Metall.* 1999, 42(4), 301-311.

[8]		Shima, S. Int. *J. Powder Metall.* 2002, 38(2), 41-47.

[9]		Dietze, M.; Buchkremer, H.P.; Stöver, D. *Metal Powder Report*. 1991 October, 30-35.

[10]		Li, W.B.; Haggblad, H.A. *Powder Metall.* 1997, 40(4), 279-281.

[11]		Kim, K.T.; Suh, *J. Powder Metall.* 1990, 33(1), 40-44.

[12]		Kovalchenko, M.S. Sov. *Powder Metall. Met. Ceram.* 1991, 29(9), 753-756.

[13]		Kakar, A.K.; Chaklader, A.C.D. *J. Appl. Phys.* 1967, 38(8), 3223-3230.

[14]		Laptev, A.M.; Ul'yanov, A.N. Sov. *Powder Metall. Met. Ceram.* 1984, 23(3), 183-186.

[15]		Early, J.G.; Lenel, F.V.; Ansell, G.S., *Trans. Met. Soc.* 1964, 230, 1641-1645.

[16]		McClelland, J.D. J. Amer. *Ceramic Soc.* 1964, 44(10), 47-48.

[17]		Spriggs, R.M.; Vasilos, T. J. Amer. *Ceramic Soc. AIME* 1961, 40(4), 187-189.

[18]		Helle, A.S.; Easterling, K.E.; Ashby, M.F. *Acta Metall.* 1985, 33(12), 2163-2174.

[19]		Gilman, P.S.; Gessinger, G.H. *Powder Metall. Int.* 1980, 12(1), 38-40.

[20]		Fischmeister, H.F.; Artz, D.E. *Powder Metall.* 1983, 26(2), 82-88.

[21]		Smith, D.W.; Marth, T., In Modern Developments in Powder Metallurgy; Hausner, H.H.; Antes, H.W.; Smith, G.D.; Principles and Processes; *Metal Powder Industries Federation*: Princeton, NJ, 1981; Vol. 12, pp 835-854.

[22]		Meyer, B.A.; Smith, D.W. *Industrial Engineering Chemistry Fundamentals* 1985, 24, 360-368.

[23]		Smith, D.W.; Smugeresky, J.E.; Meyer, B.A. *Technical Report SAND87-8227, Sandia National Laboratories*: Livermore, CA, 1987; October.

[24]		Exner, E.; Hougardy, P. Quantitative Image Analysis of Microstructures; DGM *Informationsgesellschaft* mbH.: Berlin, Germany, 1988, pp 10-30.

[25]		Rabin, S. A.I. *Game programming wisdom*, Charles River Media: Princeton, NJ, 2002.

[26]		Balshin, M. Yu. Vestnik. *Metalloprom.* 1938, 18(16), 124-137.

[27]		Heckel, R.W. *Trans. Metall. Soc.* 1961, 221, 1001-1008.

[28]		Kawakita, A.; Ludde, K.H. *Powder Technol.* 1970, 4, 61-68.

[29]		GE, R.D., *Powder Metall. Sci. Technol.* 1995, 6(3), 20-24.

[30]		Panelli, R.; Filho, F.A. *Powder Metall.* 1998, 41(2), 131-133.

[31]		Secondi, *J. Powder Metall.* 2002, 45(3), 213-217.

[32]		Park, H.; Kim, K.T. Mat. *Sci. Eng. A.* 2001, 299, 116-124.

[33]		Okimoto, K.; Oyane, M.; Shima, S. *J. Jpn. Soc. Powder Powder Metall.* 1975, 22(6), 177-184.

[34]		Bruhns, O.; Sluzalec, A. *Int. J. Mech. Sci.* 1993, 35(9), 731-740.

[35]		Evans, U.R. The corrosion and oxidation of metals: first supplementary volume; Edward Arnold Ltd., London, 1968.

[36]		Taylor, G. F. Apparatus for Making Hard Metal Compositions, United States Patent No. 1896854, 1933; February 7.

[37]		Lenel, F.V. J. *Metals*, 1955, 7(I), 158-167.

[38]		Suzuki, T.; Saito, S. *J. Jpn. Soc. Powder Powder Metall.* 1971, 18(1), 28-32.

[39]		Saito, S.; Ishiyama, T.; Sawaoka, A. *Bull. Tokyo Inst. Tech.* 1974, 120, 137-140.

[40]		Hara, Z.; Akechi, K. *Titanium'80* 1982, 2265-2274.

[41] Okazaki, K. *Rev. Particulate Mat.* 1994, 2, 215-269.

[42] Istomina, T.I.; Baidenko, A.A.; Raichenko, A.I.; Goldberg, M.A.; Svechkov, A.V. *Sov. Powder Metal. Met. Ceram.* 1983, 22(11), 957-960.

[43] Burenkov, G.L.; Raichenko, A.I.; Suraeva, M. *Sov. Powder Metall. Met. Ceram.* 1987, 26(9), 709-712.

[44] SUKHOV, O.V.; Baidenko, A.A.; Istomina, T.I.; Raichenko, A.I.; Popov, V.P.; Svechkov, A.V.; Goldberg, M.A. *Sov. Powder Metall. Met. Ceram.* 1987, 26(7), 530-532.

[45] Yokota, M.; Nagae, T.; Nose, M. 1998 PM *World Congress Sintering*, Granada, Spain, 1998, 284-289.

[46] Moriguchi, H.; Tsuduki, K.; Ikegaya, A. *Powder Metall.* 2000, 43(1), 17-19.

[47] Albella, J.M.; Martínez, J.M. *Física de dieléctricos*, Marcombo: Barcelona, SP, 1984; p 143.

[48] Montes, J.M.; Rodríguez, J.A.; Herrera, E.J. *Rev. Metalurgia Madrid* 2003, 39, 99-106.

[49] Montes, J.M. Ph. Thesis: Modelling of resistance sintering under pressure of metallic powders (in Spanish); UNED, June 2004.

In: Trends in Materials Science Research
Editor: B.M. Caruta, pp. 191-217

ISBN: 1-59454-367-4
© 2006 Nova Science Publishers, Inc.

Chapter 8

BARIUM ION LEACHING AND ITS EFFECT ON AQUEOUS BARIUM TITANATE TAPE PROPERTIES

Dang-Hyok Yoon[1] and Burtrand I. Lee[2]
School of Materials Science and Engineering, Olin Hall, 340971,
Clemson University, Clemson, SC 29634-0971, USA

Abstract

In this chapter we present three processing aspects to meet the challenges in MLCC processing in aqueous media. One is the understanding of Ba^{2+} ion leaching behavior in water which is known to be one of the drawbacks of water-based slip systems. The second is the effect of excess Ba^{2+} ion on tape properties by using an external Ba^{2+} ion source. The third is the potential solution for the reduction of the amount of Ba^{2+} leaching by using a polymeric passivation agent layer (PAL). EDTA titration method was shown to be in determining the amount of Ba^{2+} ion leaching from $BaTiO_3$ in water. The greater extent and the faster rate of Ba^{2+} leaching were found at the lower solution pH. The excess free barium ions expressed by means of the Ba/Ti ratio adversely affected most tape properties. Increase in the slip viscosity, porosity, agglomeration, and along with a decrease in mechanical properties and green/sintered density were found with the increase in the Ba/Ti ratio. An effort was made to correlate these phenomena with Ba^{2+} leaching in water for realistic MLCC applications. To passivate $BaTiO_3$ surface from Ba^{2+} ion leaching, PAL was formed by drying the slurry after adding a commercial polymeric dispersant. Compared to the conventional dispersant adding method, this PAL method was more effective in reducing the amount of Ba^{2+} leaching. Based on these results, we made practical recommendations for aqueous processing of $BaTiO_3$ powders.

Introduction

Barium titanate ($BaTiO_3$) is one of the most widely used ceramic raw materials in the electro-ceramic industry. Since the discovery of ferroelectricity in $BaTiO_3$ in the 1940s [1], much research has been carried out on this ceramic material. Nowadays, multilayer ceramic

---

[1] E-mail address: dhyoon@yumail.ac.kr, Currently at Yeungnam University, Korea.
[2] E-mail address: burt.lee@ces.clemson.edu

capacitors (MLCCs) based on $BaTiO_3$ are one of the most important electronic components in surface mounted electronic circuits. Capacitor applications in electronic equipment include discharge of stored energy, blockage of direct current, coupling of circuit components, by-passing an AC signal, frequency discrimination, and transient voltage and arc suppression. Common electronic appliances contain many MLCCs; a typical watch contains 2 - 4, a video camera or cell phone 250, a laptop computer 400, and an automobile over 1000. MLCCs having the highest volumetric efficiency are currently less than 3 μm of the dielectric layer thickness and are comprised of several hundreds of layers [2]. The dielectric layers in MLCCs are produced from tape casting of a $BaTiO_3$ slip. It is also expected that the minimum dielectric thickness will be less than 1 μm within several years.

Slips for MLCCs are prepared by dispersion of ceramic powders in a liquid medium with processing additives, such as binders, dispersants, plasticizers, etc. Depending on the liquid medium used to dissolve the binder and to suspend the ceramic particles, slip systems can be divided into water- or solvent-based systems. Water-based slip systems are known to have problems such as greater tendency for foaming, and a slower drying rate compared to a solvent-based one [3-5]. The other problem is Ba^{2+} ion leaching from the $BaTiO_3$ surface in water [6-9], resulting in variations of the slip viscosity, Ba/Ti ratio of the powder, sintering density, and dielectric properties [10]. In addition, leached Ba^{2+} from the $BaTiO_3$ surface can interact in an adverse manner with various polymeric additives [11,12]. The Ba^{2+} ions present in the liquid medium can cause the folding of the molecular chains of the polymeric additives caused by a salting-out mechanism. This folding will impair not only the homogeneous dispersion of $BaTiO_3$ particles but also lead to segregation creating of defective holes in the fired tapes as a result of the burnout of the salted-out polymers. In order to use the water-based slip system in the MLCC industry with greater reliability, therefore, control or reduction of the Ba^{2+} leaching is essential, not only to preserve the chemical homogeneity in the powder compact but also to prevent adverse interactions with such organic species as dispersants and binders. Due to these problems of water-based slip system, organic solvents have been used traditionally in MLCC industry as the liquid media [11]. In spite of many advantages over a water-based slip system, a solvent-based slip system has its own drawbacks such as flammability, toxicity, and high cost. Therefore, slips using water are still desired and are worthwhile to be investigated.

One possible concept to reduce the amount of Ba^{2+} leaching in water is the creation of a passivation agent layer (PAL) which covers the $BaTiO_3$ surface and acts as a surface diffusion barrier. It has been reported [11] that oxalic acid PAL can be used for the prevention of Ba^{2+} leaching in water considering the fact that an insoluble alkaline earth oxalate is easily formed on $BaTiO_3$ surface. The PAL, oxalic acid, was also applied to yttria-stabilized zirconia system [13]. It was hypothesized that the oxalate formed on the surface of both powders remained and acted as a diffusion barrier. However, it is doubtful whether this oxalate layer will still remain at the powder surface under a high-energy milling process which is widely performed for powder dispersion. Moreover, once this insoluble barium oxalate is formed and precipitated from the reaction between oxalic acid and Ba^{2+}, it is difficult to determine precisely the amount of Ba^{2+} leaching in solution because most of the quantitative analysis methods use a clear supernatant [6-8,14].

Although $BaTiO_3$ has been extensively studied, little information has been published on the whole slip formulation for MLCC application due to probably the highly competitive ceramic industry or complexity of the slip system, especially with water-based slip system. Hence, the main aim of this chapter is to illustrate to improve the aqueous processing of $BaTiO_3$ slips for MLCC applications via the following three experimentations. The first is to understand the Ba^{2+} ion leaching behavior from $BaTiO_3$ in water under different conditions and the effect of surface $BaCO_3$ on Ba^{2+} ion leaching. The second is to study the effect of excess Ba^{2+} ion on tape properties in terms of slip viscosity, green and sintered properties of $BaTiO_3$ bodies prepared from different Ba/Ti ratios of the slips by using an external Ba^{2+} ion source. The third is to investigate the effects of PAL on reducing the amount of Ba^{2+} ion leaching in water using polymeric dispersants. By following several characteristic steps of the actual MLCC production processes including slip preparation, tape casting, K-square (small rectangular-shaped chips) preparation and sintering, output responses such as slip, green, and sintered properties are compared in this experiment. Here slurry is defined as a ceramic suspension with particles in a liquid medium with a dispersant, and a slip as ceramic particles in a liquid medium with all the processing additives including binder, dispersant, plasticizer, surfactant, and defoamer.

Experimental

Ceramic Powder

The ceramic powder used in this study was a hydrothermally prepared BT-8 (Cabot Performance Materials, Boyertown, PA) with a mean particle size of 0.24 μm, a specific surface area of 8.50 m^2/g, and a Ba/Ti ratio of 0.998. According to supplier's data, BT-8 is high purity $BaTiO_3$ powder with small amount of impurities such as C (649 ppm), Cl (231 ppm), and Fe (58 ppm). Figure 1 shows the morphology of BT-8.

Figure 1. SEM micrograph of BT-8 starting ceramic powder.

Barium Ion Leaching in Water

To understand Ba^{2+} ion leaching from $BaTiO_3$ powder surface in water at different conditions, the following sets of experiments were performed.

As the first step, ensuring the reliability measuring the amount of Ba^{2+} in water, an exactly known amount of Ba-acetate was dissolved in water, followed by the EDTA titration. The effect of any organic additives in this blank test was checked by using oxalic acid and by adding 0.5 wt% dispersants with respect to BaTiO$_3$ powder. The solution was titrated with a 0.001 M EDTA solution using a Mettler Titration Meter (DL50 Titrator, Mettler Toledo, Switzerland).

In the second step, to determine the amount of Ba^{2+} ions leaching from the BaTiO$_3$ particles, 6 g of BT-8 powder was added to 24 ml of deionized water. This mixture was then stirred mildly by hand with a spatula for 1 minute to disperse the particles. NH$_4$OH and CH$_3$COOH were used to adjust the solution pH by using an Accumet 925 pH/ion Meter (Fisher Scientific, Pittsburgh, PA). The solution pH was readjusted after an equilibration period of 5 hours, if drift from the initial pH. After being aged for the desired length of times in air, the slurry of BaTiO$_3$ was centrifuged at room temperature at 13000 rpm for 30 min. A 12 ml aliquote of the supernatant was diluted to 50 ml with deionized water, and the pH was readjusted to 12 by a 10 M KOH solution for EDTA titration.

The third investigation was the effect of BaTiO$_3$ calcination on the amount of Ba^{2+} leaching. After treating the BaTiO$_3$ powder at temperatures of 400°, 600° and 1000°C for 2 hours in air to remove surface adsorbed species, Fourier transform infrared (FT-IR, Magna FT-IR spectrometer, Model 560, Nicolet Instrument) spectra of each sample were immediately taken to prevent re-adsorption of gaseous contaminants on the BaTiO$_3$ surface. Spectral scans were made in air using a diffuse reflectance stage with a resolution of 4 cm^{-1} after taking a background calibration to subtract the background noise.

Excess Barium Ion Effects on Tape Properties

Using the same hydrothermal BT-8 powder, the effects of intentionally added excess barium ions, in terms of Ba/Ti ratio, on tape properties were investigated. Six kinds of slips with different Ba/Ti ratios (Ba/Ti = 0.998, 1.005, 1.010, 1.020, 1.030, 1.040) were prepared by adding Ba(OH)$_2$·8H$_2$O to minimize any anionic effects. Slips for tape casting were prepared with an acrylic formulation binder solution (WB40B, Polymer Innovations, San Marcos, CA), 60 g of BaTiO$_3$ powder, and other processing additives. The amount of acrylic binder resin added was 8 wt. % with respect to the BaTiO$_3$ powder, and dispersant in 0.5 wt. % in the slip. Additionally, surfactant (S465, Air Products and Chemicals, Allentown, PA) in 0.5 wt. % and defoamer (DF001, Polymer Innovations, San Marcos, CA) in 0.2 wt. % with respect to the total slip were added. The slip was adjusted to pH of 9.7 ± 0.1 by using ammonium hydroxide, and was subjected to 24 hours ball-milling using yittria-stabilized zirconia media. To remove any residual air bubbles and to stabilize the slip, 10 rpm slow rolling for 24 hours without the media was performed. The rheological behavior of the slips was characterized using a programmable rheometer (Model DV-III, Brookfield, Stoughton, MA). The measurements utilized a small sample adapter and a SC4-21 spindle at a constant temperature of 25°C in the shear rate range of 1 to 140s^{-1} in ascending and descending order.

After producing green tapes by means of tape casting on moving polypropylene film using a tape caster (TTC-1000, Richard E. Mistler, Morrisville, PA) with a 4-inch wide single doctor blade, mechanical properties of these tapes were tested using a tensile tester (Minimat 2000, Rheometric Scentific, Piscataway, NJ) with a 200 N load cell. Dumbbell-shaped specimens with 3 mm width and 10 mm length were cut from the green tapes toward casting direction using a sample cutter. The tensile tests were performed at ambient temperature of 22 ± 1°C with the pulling rate of 1 mm/min. At least five samples were tested to calculate the average tensile properties of these tapes. A Windows-supported software was used to control the instrument and to gather the stress-strain data. After gold coating, the green microstructure of tapes was observed using a scanning electron microscope (SEM; Hitachi S-3500N). Six K-squares with the thickness of approximately 1 mm were made from each green tape by stacking, laminating at 300 kPa for 5 min at 80°C, and cutting into small rectangular shape (2.0×1.8 cm^2). However, the green tapes with Ba/Ti ratio of 1.030 and 1.040 could not be laminated due to their poor tackiness, and therefore, they were not able to process further. Green density of the K-squares was determined using a geometric method. The thickness and length were measured using a micrometer accurate to 0.001 mm and the weight was measured using a scale accurate to 0.1 mg. Sintering of K-squares was performed at 1280 and 1320°C, and with each layer of green tape at 1250°C for 2 hours in air with a heating and cooling rate of 3°C/min. Sintered density of K-squares was measured using the same geometric method. K-squares were electroded with silver paste and then fired at 550°C for 12 minutes to burn-out the polymeric species in the silver paste. After a gold coating, the sintered microstructure of the K-square surface was examined under SEM. Finally, dielectric permittivities and dissipation factors of K-squares were measured at 1 kHz/1 V$_{rms}$, using a HP 4284A LCR meter after stabilizing the K-squares for 6 hour after electrode firing. All the slip composition including chemicals used and experimental procedures were exactly same as those of the optimum condition for water-soluble acrylic binder system shown in our previous reports [15,16].

Surface Passivation Using Polymeric Dispersants

The effects of passivation agent layer (PAL) on the amount of Ba^{2+} leaching using commercial dispersants on aqueous tape properties were investigated. In general, a dispersant is added to a liquid medium directly to enhance the slurry dispersion. After adding dispersant, however, some slurries were dried to produce a PAL which adsorbed on the powder surface acting as a polymeric diffusion barrier for Ba^{2+} leaching in this experiment. The aim of this study is to investigate the differences in Ba^{2+} dissolution, slurry, slip, and green tape properties between conventional dispersant method and PAL mode.

In the first step, aqueous slurries of 55 wt. % BaTiO$_3$ with different dispersant concentrations were prepared to decide the adequate amount of each dispersant for monolayer adsorption. The slurries were milled for 2 hours and stabilized for 24 hours without agitation in air atmosphere. To acquire a clear supernatant, slurries were centrifuged at a speed of 13000 rpm for 20 minutes. The adsorption isotherms of the PALs were determined by a gravimetric method after drying the supernatant. Based on these results, the amount of dispersant in this experiment was fixed at 0.5 wt. % with respect to the ceramic powder.

As the second test, a PAL-coated powder was made by adding three different kinds of dispersants. After dispersing the powder in water containing a dispersant, this slurry was immediately dried at 120°C for overnight to minimize the Ba^{2+} leaching at this step. To maintain the overall Ba/Ti ratio constant in the slurry, no portion of the liquid was discarded but removed by drying. Two kinds of aqueous slurries of 55 wt. % BaTiO$_3$ with 0.5 wt% of dispersant were prepared; dispersant and PAL modes. The rheological behaviors of the slurries were compared for the two different modes. EDTA titration was performed with the supernatant obtained from the slurries to measure the amount of Ba^{2+} leaching after 24 hours ball-milling and 1-day aging at pH=8.

As the third test, slips were prepared with an acrylic formulation binder solution with the same manner described above. Dispersant in 0.5 wt. % was added to the dispersant mode slip, but PAL-coated powder was directly used without adding any dispersant because the powder already contained the dispersant. After producing green tapes, the tape surface morphologies were compared under SEM. K-square preparation and green density measurements were performed. All the materials used are listed in Table 1. The schematics of the tape preparation are given in Figure 2.

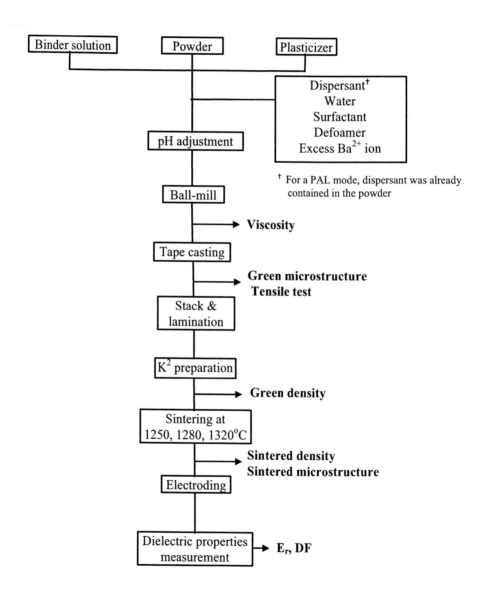

Figure 2. Schematic experimental procedure for tape preparation and characterization.

Table I. Materials information

Material	Name	Description
Binder solution	WB40B	formulated acrylic binder
		$T_g = 40^{\circ}C$, resin MW = 100,000
Dispersant	PMMA/PEG	modified polymethylmethacrylate, MW = 35,000
	APA	ammonium salt of polyacrylic acid, MW = 6,000
	PApA-Na	sodium salt of polyaspartic acid, MW = 4,500
Plasticizer	PL001	polyether polyol
Surfactant	S465	tetramethyldecynediol + ethylene oxide
Defoamer	DF001	modified silicone copolymer
Oxalic acid		$(COOH)_2 \cdot 2H_2O$

Results and Discussions

Barium Ion Leaching in Water

Figure 3 shows the results of the blank test with the different organic species. The y-axis represents the percent difference between the actual amount of Ba^{2+} and the titrated value. As the data show, the titrated values of the 4 samples agree with the actual amount of Ba^{2+}; only oxalic acid shows a difference. With oxalic acid added sample, precipitates were observed during the titration process. It is well-known that alkaline earth oxalates are quite insoluble and interfere with the volumetric determination by EDTA [17]. As a result, we can conclude that the EDTA titration method can be used to detect the amount of dissolved Ba^{2+} ions accurately unless a precipitate is formed.

Figure 4 shows the amount of Ba^{2+} leaching and the resultant Ba/Ti molar ratios of the starting $BaTiO_3$ powder calculated after being aged for 1 and 2 days in various pH solutions. The amount of Ba^{2+} leaching shows a significant dependency on the solution pH, i.e., the higher leaching amount and the faster leaching rate at the lower pH. Even though the system is not in equilibrium condition, this pH-dependent Ba^{2+} leaching behavior shows the similar trend of the equilibrium condition which was reported by Lencka and Riman [18]. As explained by the theoretical model [18], it was found that there is approximate linear log relationship between $[Ba^{2+}]$ and pH.

However, the situation is complicated in normal environment because of the general presence of carbon dioxide in water. The reaction of $BaTiO_3$ with CO_2 dissolved in water leads $BaCO_3$ formation [18]. This $BaCO_3$ increases the Ba^{2+} concentration in the solution by dissolving according to the solubility product of $BaCO_3$. As a result, one can expect an appreciable Ba^{2+} leaching in water unless the pCO_2 is extremely low. In contrast, $TiO_2(s)$ is generally regarded as insoluble in water. [19,20] Therefore it is likely that more Ba^{2+} will be released from the $BaTiO_3$ surface until the surface becomes a "TiO_2-like" character.

Figure 5 shows the Ba^{2+} leaching behavior at pH=8 and at room temperature as a function of aging time since the usual industrial practice is in the pH of 8 to 10. The amount of Ba^{2+} leaching for the first 30 minutes could not be measured due to the time needed for the sample preparation. As shown in Figure 5, the amount of Ba^{2+} leaching is quite high for the first measurement at 30 minutes of aging. After 30 minutes, the leaching rate gradually decreases reaching a plateau at a time greater than 100 hours.

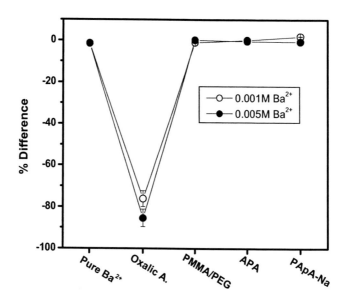

Figure 3. Results of the EDTA blank test with Ba-acetate solution after adding different organic material in water.

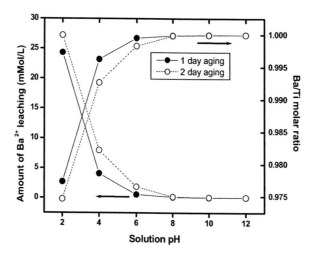

Figure 4. The amount of Ba^{2+} leaching and the resultant Ba/Ti molar ratio as a function of the solution pH without organic material.

The results on the time-dependent behavior of Ba^{2+} leaching in water reported by others [7,9,14] vary markedly. Neubrand et al. [14] observed an instantaneous increase of Ba^{2+} leaching in water within the first few minutes and then a gradual decrease at a pH of approximately 10. They explained this finding as a re-adsorption of Ba^{2+} ions onto the hydrated $BaTiO_3$ surface. On the other hand, Blanco-Lopez et al. [7] observed a gradual increase in the amount of Ba^{2+} leaching in water at pH of 4 within 5 hours of aging, after an instantaneous increase at the beginning which agrees with our results. Utech [9] explained that Ba^{2+} leaching in water at pH 8, 10 and 12 occurred instantaneously, reaching a steady-state within 5 minutes, i.e., no Ba^{2+} concentration change after 5 minutes. Regarding the amount of Ba^{2+} leaching, the average leached Ba^{2+} concentration in our experiment lies between the results of Neubrand et al.'s [14] and Utech's [9]. The variations observed here must be caused by the difference in the state of $BaTiO_3$ surface of each investigator used for the experiment. The degree of surface carbonation of $BaTiO_3$ can vary widely [21,23].

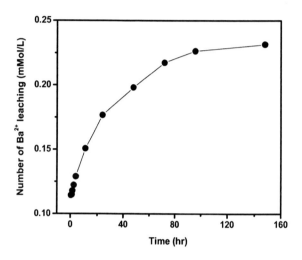

Figure 5. The amount of Ba^{2+} leaching vs. time at pH=8 and at room temperature.

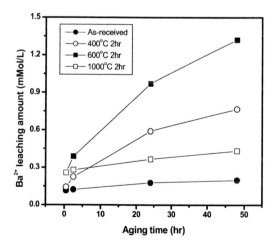

Figure 6. Effect of heat treatment of $BaTiO_3$ powder on Ba^{2+} leaching rate at room temperature.

Pre-heat treatment of $BaTiO_3$ in air increased the amount of Ba^{2+} leaching as shown in Figure 6 with the highest amount of Ba^{2+} leaching at 600°C and then a decrease at 1000°C. This result can not be explained by the specific surface area change which is shown in Figure 7. By increasing pre-heat treatment temperature, specific surface area is decreased, and particle size is increased. To check the next possible variable, the effect of surface $BaCO_3$ on Ba^{2+} leaching, FT-IR spectra of the pre-heated $BaTiO_3$ were used. The FT-IR spectra of these powders measured within 10 minutes of the heat treatment shown in Figure 8 indicate band height differences at approximately 3400 and 1200 – 1500 cm^{-1}. According to previous research [21-23], the broad IR bands at approximately 3383 cm^{-1} are attributed to surface adsorbed water, 1426 and 1469 cm^{-1} to two different carbonates: the former to lattice carbonate and the latter to surface carbonate, and 785 and 475 cm^{-1} to $BaTiO_3$. Small peaks at 1750 and 2450 cm^{-1} are also assigned to carbonate species [22]. Semi-quantitative comparison of the species by band intensity was calculated using the ratios of the band intensity caused by the species divided by the band intensity of $BaTiO_3$ at 785 cm^{-1}. The increasing OH band height, peaking at 600°C, can be explained by the diffusion of the hydroxyl ion (OH) to the $BaTiO_3$ surface which had been incorporated into the lattice during the hydrothermal synthesis of this powder. According to Hennings et al. [24], there is an appreciable amount of lattice-incorporated protons and hydroxyl ions (≈ 0.4mol/mol $BaTiO_3$) present in hydrothermally synthesized $BaTiO_3$ powder. As they found, this OH diffuses to the $BaTiO_3$ surface during the calcinations through 800°C [24].

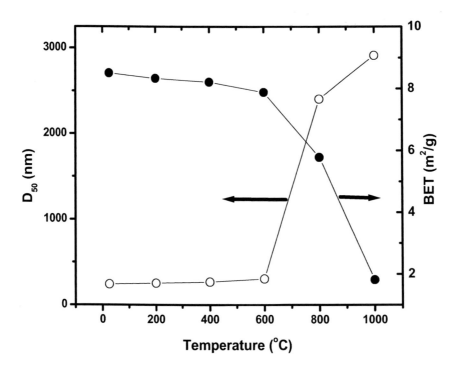

Figure 7. Specific surface area and median particle size changes of $BaTiO_3$ powder with pre-heat treatment temperature.

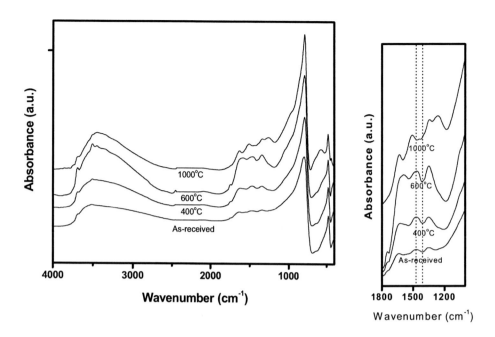

Figure 8. FT-IR results of BT-8 after calcining at various temperatures and enlarged spectra between 1000 – 1800 cm^{-1}.

For the carbonate peaks shown in Figure 8, the band height ratios at 1426 and 1469 cm^{-1} were calculated using the baseline method suggested by Brezinski [25]. These carbonate band height ratios were obtained after normalizing the carbonate band heights by that of BaTiO$_3$ at 785 cm^{-1}. Both of the lattice and surface carbonate band height ratios at 1426 and 1469 cm^{-1} increased, peaking at 600°C, and then decreased at 1000°C as shown in Figure 9 (a). This means that the carbonates at the surface decomposed at temperature above 600°C. Both of the small carbonate peaks at 1750 and 2450 cm^{-1} also agree with the 1426 and 1469 cm^{-1} carbonates. These results also agree with the Ba^{2+} leaching behavior shown in Figure 6. This behavior is further demonstrated by the linear relationship between the carbonate band height ratios at 1469 cm^{-1} and the amount of Ba^{2+} leaching shown in Figure 9 (b). This finding indicates that Ba^{2+} leaching depends more on the amount of surface carbonate than on the diffusion of the lattice Ba^{2+} for the pre-heat treated BaTiO$_3$ powders. As the temperature increases up to 600°C, the BaTiO$_3$ surface in contact with the atmospheric CO$_2$ and H$_2$O forms carbonates which are fairly soluble in water. Evidently, this form of carbonate has no passivating capability for Ba^{2+} leaching.

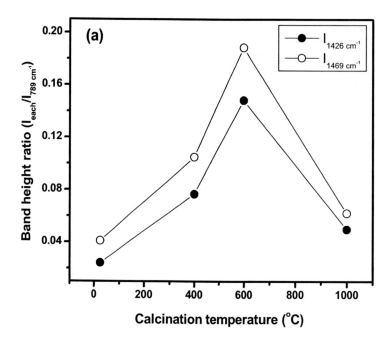

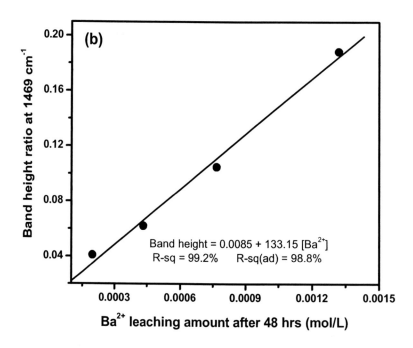

Figure 9. (a) Normalized carbonate band height ratios at 1426 and 1469 cm^{-1} after heat-treatment of BaTiO$_3$ at different temperature in air for 2 hours and (b) the linear relationship between the normalized carbonate band height ratios and the amount of Ba^{2+} leaching after 48 hours aging.

Excess Barium Ion Effects on Tape Properties

Figure 10 shows the viscosity of slips as a function of the Ba/Ti ratio. The slip viscosity decreases slightly up to Ba/Ti ratio of 1.010, then increases drastically in the range of 1.010 – 1.020. A concentration of barium ion, Ba/Ti>1.020, shows a gradual increase in the viscosity beyond the sharp increase at Ba/Ti=1.010 – 1.020.

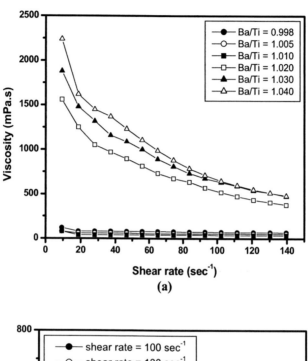

(a)

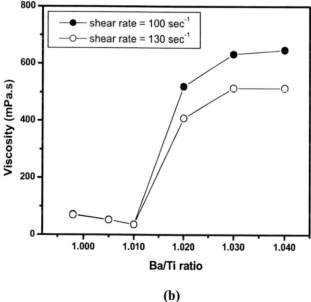

(b)

Figure 10. Viscosity of the slips with different Ba/Ti ratios (a) for overall shear rates, and (b) for fixed shear rate.

Figure 11 presents the surface morphology of a green tape. The tape without any excess Ba^{2+} ion (Ba/Ti = 0.998) shows a dense and smooth surface which agrees with our previous reports [15,16]. There is no apparent difference between Ba/Ti ratio of 0.998 and 1.005, while small pores start to appear from Ba/Ti ratio of 1.010. For Ba/Ti ratio of more than 1.020, the green surface shows a heavily agglomerated and rough morphology.

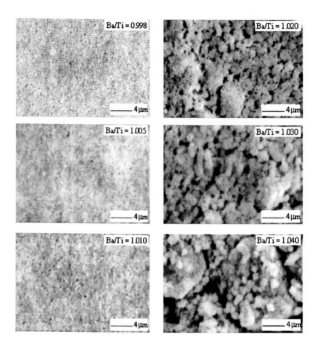

Figure 11. SEM microstructure of top surface of green tapes with different Ba/Ti ratios.

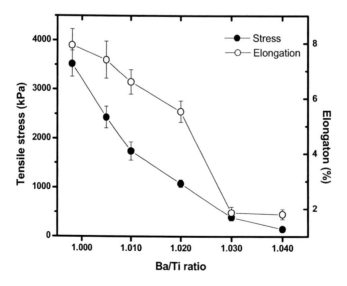

Figure 12. Engineering tensile strength and strain of the green tapes with different Ba/Ti ratios.

Figure 12 shows the engineering tensile strength and the % elongation up to the fracture of the green tapes. Both of the properties decrease with the increase in Ba/Ti ratio. The tapes with the Ba/Ti ratio of 1.030 and 1.040 show not only poor mechanical properties but also little tackiness. Due to this poor tackiness, it was impossible to laminate these tapes to make K-squares for further tests.

The green density of K-squares also decreases with the increase of Ba/Ti ratio as shown in Figure 13. This gradual decrease in green density agrees with the green morphological behavior shown in Figure 11. Lower green density values with the Ba/Ti ratio of 1.030 and 1.040 are expected from the porous green morphology shown in Figure 11, although K-squares were not available for these compositions.

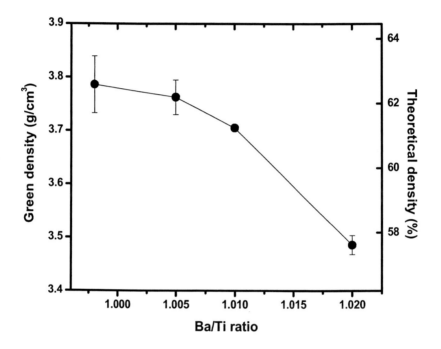

Figure 13. Green density of K-squares with different Ba/Ti ratios.

The overall change in the slip viscosity with Ba/Ti ratio variation means an interaction between Ba^{2+} and polymeric or organic additives in the slip. This result clearly does not come from the effect of hydroxyl ions of $Ba(OH)_2 \cdot 8H_2O$ because the pH of the all slips were kept constant at 9.7 using NH_4OH. Due to the multivalent positive charge of barium ions, the interactions between Ba^{2+} and some anionic polymeric species are expected. Li and Jean [26] reported that the dissolved free Ba^{2+} ions caused flocculation with an anionic ammonium salt of poly acrylic acid dispersant by the formation of physical complex, thus degrading its dispersing effectiveness in aqueous $BaTiO_3$ suspensions. Based on the behavior of slip viscosity and green body properties by the excess free Ba^{2+} ions in terms of Ba/Ti ratio, it is clear that the excess Ba^{2+} deteriorates the dispersion of the slips. However, one of the more important chemical species which reacts significantly with Ba^{2+}, and thus producing a worse situation, may be the plasticizer polyether polyol. The role of a plasticizer in a slip is to confer sufficient flexibility and tackiness to the green tape for easy handling and lamination [27]. Due to the many hydroxyl functional groups on polyether polyol, it is easy to form coordination bonds with Ba^{2+} ion. By the formation of this chemical complex, both of the mechanical properties and tackiness are decreased with Ba/Ti ratio increase, especially with the Ba/Ti ratio of more than 1.030 as shown in Figure 12.

Figure 14 shows the microstructure of each tape layer sintered at 1250°C for 2 hours. Fine grains are distributed uniformly with the Ba/Ti ratio of 0.998, while the tape with Ba/Ti ratio of 1.005 shows the largest grain size. With the Ba/Ti ratio of more than 1.030, sintered tapes show much porous morphology. The sintered K-square surface morphologies at 1280 and 1320°C are presented in Figure 15. The K-squares with the Ba/Ti ratio of 0.998 and 1.005 show an abnormal grain growth at 1280°C and fully grown microstructures at 1320°C with the average grain size of more than 50 µm. On the other hand, the microstructure of K-squares with Ba/Ti ratios of more than 1.010 does not show any significant change as the sintering temperature increased. The average grain size at 1320°C sintering for these compositions remains approximately 3 µm which is quite smaller than those of Ba/Ti = 0.998 and 1.005.

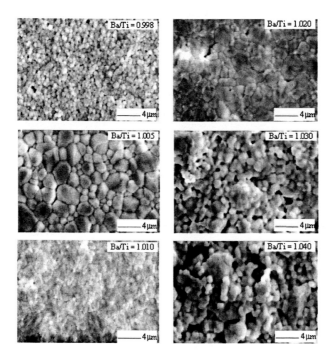

Figure 14. Microstructure of sintered tape surface at 1250°C for 2 hours in air.

The sintered densities of the K-squares, as shown in Figure 16, are in the range of 88 – 95% of theoretical $BaTiO_3$ density. The K-squares sintered at 1320°C show higher value than those sintered at 1280°C in general. Sintered density is the highest at Ba/Ti ratio of 1.005 and then decreases with Ba/Ti ratio increase.

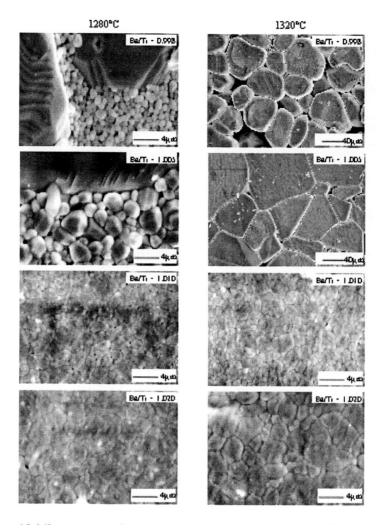

Figure 15. Microstructure of sintered K-square surface at 1280 and 1320°C for 2 hours.

The tapes with Ba/Ti ratio of 1.005 show the highest sintered density and the largest grain size among the samples. Although BaTiO₃ ceramic with a composition of slight excess in Ti was reported to yield a higher sintered density [80], the same trend seems to have occurred here for a composition of slight excess in Ba^{2+}. The similar results were reported by Hu et al. [28] with the Ba/Ti ratio of 1.001 and 1.002. With the Ba/Ti ratio of greater than 1.010, there is no abnormal grain growth even at 1320°C. The grain size is shown to decrease with the Ba/Ti ratio increase, which agrees with other reports [28,29]. Hu et al. [28] also proposed that the fine uniform sintered microstructure with a high Ba/Ti ratio results from the suppression of grain growth by the secondary Ba_2TiO_4 phase. Due to the low solid solubility limit in $BaTiO_3$ [30], Ba_2TiO_4 phase is formed easily and located along the grain boundaries.

Figure 17 shows the dielectric permittivity and dissipation factor of K-squares sintered at 1280 and 1320°C with different Ba/Ti ratios. Both of the properties increase with increasing Ba/Ti ratio. The dielectric permittivities of K-squares sintered at 1280°C show higher values than those sintered at 1320°C up to Ba/Ti ratio of 1.010, while the opposite is true for Ba/Ti ratio of 1.020. In general, dielectric permittivity at room temperature is a function of sintered density and grain size for a stoichiometric BaTiO$_3$, i.e., the higher the sintered density, the higher the permittivity with the maximum permittivity peak at the grain size of around 1 μm [15,16,31-33]. However, sintered density must not be the main factor for the dielectric permittivity increase in this case due to the low sintered density of samples with high Ba/Ti ratio. Moreover, the Ba$_2$TiO$_4$ second phase itself cannot be the contributing factor due to its low dielectric permittivity. Arlt et al. [32] explained that the dielectric permittivity of stoichiometric BaTiO$_3$ ceramics is proportional to the total area of 90° domain boundaries which is originated from the equilibrium of elastic field energy and domain wall energy. According to Lee and Hong [34], high permittivity associated with the high Ba/Ti ratio is also attributed to the higher density of 90° domains related to the presence of Ba$_2$TiO$_4$ second phase. This phase increases the stress in the BaTiO$_3$ matrix because of the difference in the thermal expansion coefficients, and this high internal stress increases the density of 90° domains in BaTiO$_3$. As a whole, the permittivity behavior with Ba/Ti ratio can be explained with the combined mechanism of "grain size effect" and domain structure caused by the presence of the Ba$_2$TiO$_4$ second phase.

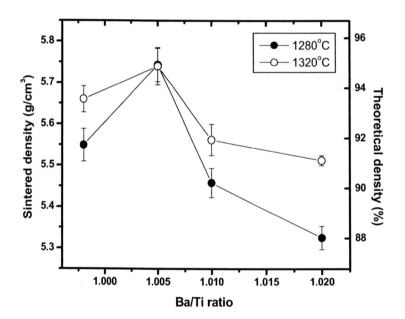

Figure 16. Sintered density of K-squares at 1280 and 1320°C with different Ba/Ti ratios.

Compared to the previous results on Ba^{2+} leaching behavior using the same ceramic powder, the actual amount of Ba^{2+} leaching after 2 days of aging is relatively small to see these differences in the practical pH condition. According to those results, the amount of Ba^{2+} leaching corresponds to the Ba/Ti ratio of 1.0002 at pH=8 and of 1.0254 at pH=2. Even though there is a difference with Ba^{2+} source between the two experiments, i.e., one from BaTiO$_3$ surface and the other from an externally added source, the results must be applicable to an aqueous MLCC processing because most of the above results are caused by free Ba^{2+} ions in the slip. The results presented in this paper clearly show the trend that free excess Ba^{2+} ions affect the slip, green and sintered properties of BaTiO$_3$ in an adverse manner. Therefore, reducing the amount of Ba^{2+} leaching in a water-based slip system is highly desired to minimize the possible adverse effects caused by Ba^{2+} ions in ensuring the reliability of final MLCC products. Moreover, due to the increasing demand nowadays for high volumetric efficiency of MLCCs with very thin layers, the importance of better dispersion and denser surface morphology are becoming more critical.

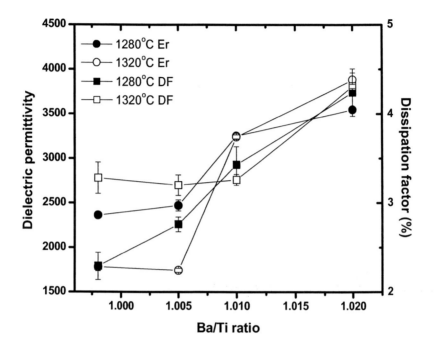

Figure 17. Dielectric permittivity (Er) and dissipation factor (DF) of K-squares sintered at 1280 and 1320°C with different Ba/Ti ratios.

Passivation Using Polymeric Dispersants

Figure 18 shows the adsorption isotherms of organic adsorbates determined by gravimetric method. PMMA/PEG shows the highest amount of adsorption and increases gradually with its concentration, which may be attributed to the high molecular weight and multi-functional groups of PMMA/PEG as explained in our previous reports [15,16]. APA and PApA-Na dispersants show an adsorption plateau at the concentration of approximately 0.5 wt%, which corresponds to the amount of dispersants for monolayer coverage. Oxalic acid shows a plateau at the concentration of 0.7 wt% which might include the effect of Ba^{2+} leached due to the formation of Ba-oxalate. Based on these adsorption isotherm results, the amount of dispersant is fixed at 0.5 wt% with respect to the ceramic powder, although PMMA/PEG and oxalic acid need slightly higher amounts for the adsorption saturation.

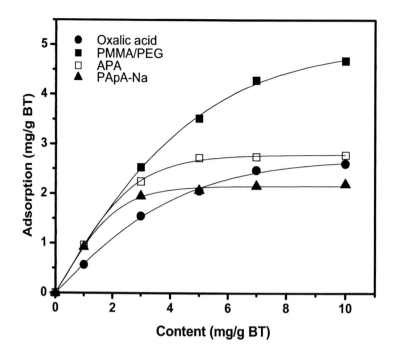

Figure 18. Adsorption isotherms of different chemicals on BaTiO$_3$ surface.

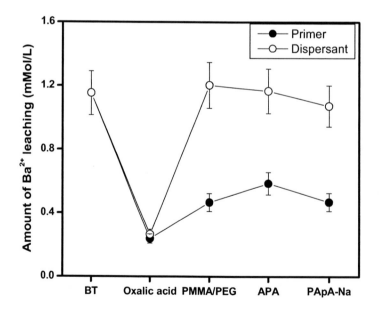

Figure 19. Comparison of the amount of Ba^{2+} leaching at pH=8 with two different adding methods of the chemicals used.

Figure 19 compares the amount of Ba^{2+} leaching between $BaTiO_3$ prepared with PALs and with dispersants modes measured by EDTA titration method. Polymeric dispersants clearly depend on the adding method for the passivation of Ba^{2+} leaching. PALs are effective in passivation, while dispersants are ineffective. More effective passivation with PAL compared to dispersant may come from the stronger binding on $BaTiO_3$ powder and more complete coverage of the surface of the ceramic powder. Oxalic acid does not show any difference with its adding method, where it shows the smallest amount of Ba^{2+} leaching among all the chemicals used. However, this low value might represent the formation of Ba-oxalate whose Ba^{2+} concentration in the supernatant could not be measured.

Figure 20 compares the viscosity of slurries of different adding methods between PAL and dispersant, where P stands for PAL and D for dispersant. There is almost no difference in viscosity between PAL and dispersant, and even slurry with PMMA/PEG PAL shows lower viscosity than that with dispersant. This means that the effectiveness of dispersion does not deteriorate by the drying process after adding PAL. Therefore, it is possible to use this PAL coating method in powder synthesis for easy dispersion. Slurry with APA shows lower viscosity than that with PMMA/PEG, which means more efficient dispersion with APA. However, when the slip is prepared by adding the binder and other additives, the slip viscosity shows different behavior as shown in Figure 21. Although there is still no difference between PAL and dispersant with the same chemicals, PMMA/PEG shows much lower slip viscosity than APA or PApA-Na. This result comes from the interaction between dispersant and binder. Therefore, it is very important to check the properties with final slip formulation, not only with slurry, for the final application because an actual MLCC process begins from the slip preparation.

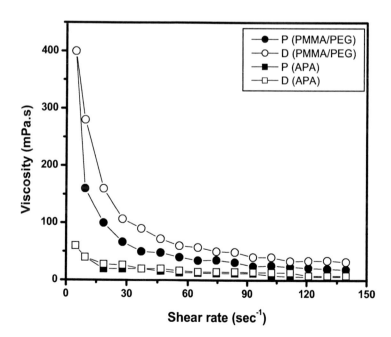

Figure 20. Comparison of slurry viscosity with two different polymeric adding methods.

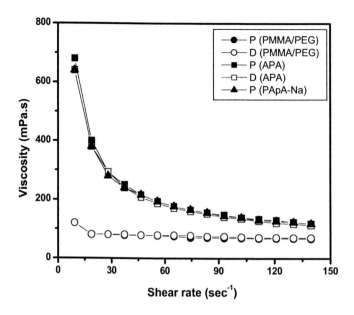

Figure 21. Comparison of slip viscosity with different polymeric species and adding methods.

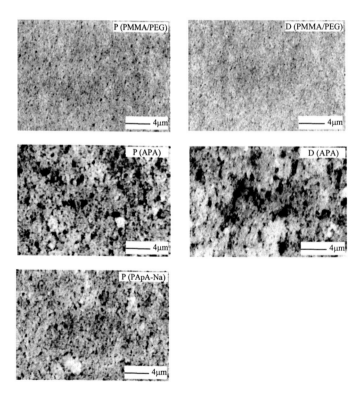

Figure 22. SEM microstructure of top surface of green tapes.

Figure 22 presents the green tape surface morphology showing a dense and smooth surface with PMMA/PEG regardless of the adding method, while tapes with APA and PApA-Na show a much more porous and rough morphology. Figure 23 shows the green density of the K-squares. The green density with PMMA/PEG is much higher than others with more than 10% difference in theoretical BaTiO$_3$ density, which agrees with the green surface morphology as shown in Figure 22. Therefore, we can conclude that the lower slip viscosity with PMMA/PEG results in the denser surface morphology and higher green density than slips with APA or PApA-Na.

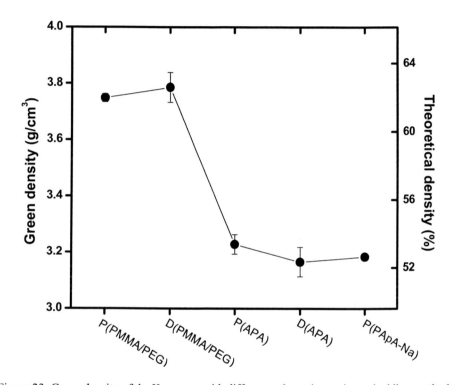

Figure 23. Green density of the K-squares with different polymeric species and adding methods.

Figure 24 shows the density of K-squares sintered at 1280 and 1320°C in air. The density values are in the range of 91.1 – 95.2% of theoretical BaTiO$_3$ density. Apparently, there is no difference between PAL and dispersant for PMMA/PEG and APA, while PApA-Na as a PAL shows the highest sintered density with the largest grain size. This is due to the effect of sodium ion in PApA-Na. The similar enhancement in sintering density with small amount of barium ion addition was reported in the previous section. Figure 25 shows the dielectric permittivity and dissipation factor for the K-squares sintered at 1280°C. In general, samples with higher dielectric permittivity show the higher dissipation factor. The K-squares with PApA-Na show relatively lower dielectric permittivity due to the largest grain size.

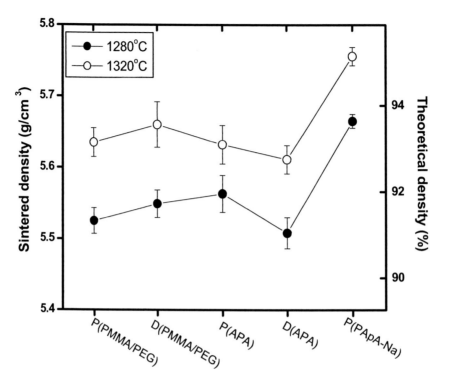

Figure 24. Sintered density of K-squares at 1280 and 1320°C.

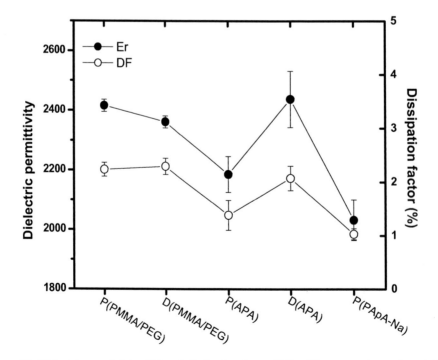

Figure 25. Dielectric permittivity (Er) and dissipation factor (DF) of K-squares sintered at 1280°C.

Although PAL is more effective than dispersant in passivation of Ba^{2+} leaching, there is no difference in other output responses such as the viscosity of slurry and slip, green morphology and density, and sintered properties. Based on the previous results, this is due to the relatively smaller amount of Ba^{2+} leaching even with dispersant to see any significant differences. However, reducing the amount of Ba^{2+} leaching is desirable using this passivation method because it was clear that the free Ba^{2+} ions adversely affect the tape properties.

Conclusions

The results presented in this paper enable the following conclusions to be drawn:

1   EDTA titration is an effective measurement method for the amount of Ba^{2+} unless a precipitate is formed as the result of the reaction of Ba^{2+} with the organic additives.
2   The amount and rate of Ba^{2+} leaching are significantly affected by solution pH; i.e., the higher leaching amount and the faster leaching rate occur at the lower solution pH, a result which can be predicted by thermodynamic calculation.
3   Pre-heat treatment of $BaTiO_3$ powder in air enhances the Ba^{2+} leaching rate due to the formation of the easily soluble carbonate which has no passivating capability for Ba^{2+} leaching.
4   Excess Ba^{2+} in aqueous $BaTiO_3$ slip deteriorates slip, green and sintered properties. It leads to poorer slip dispersion, more porous green surface morphology, lower mechanical properties and sintered density compared to the stoichiometric composition.
5   Even though the amount of Ba^{2+} leaching in a practical pH condition is relatively small to see any apparent results, reducing or prevention of the amount Ba^{2+} leaching in water is suggested for using water-based slip system in MLCC production.
6   Forming a passivation agent layer (PAL) by drying the slurry after adding a dispersant was effective in reducing the amount of Ba^{2+} leaching in water, while a conventional dispersant adding mode was less effective in passivating of Ba^{2+} leaching.
7   Using a PAL did not deteriorate any properties in terms of the viscosity of slurry and slip, green tape morphology and density, and sintered properties including a sintered density and dielectric permittivity.
8   Using a slip instead of slurry for an investigation of tape processing is required due to the interactions among the processing additives.

References

[1]   Rae, M. Chu and V. Ganine, "Barium Titanate – Past, Present and Future," pp. 1-12 in Dielectric Ceramic Materials. Ceramic Transactions, Vol. 100, Edited by K. M. Nair and A. S. Bhalla, American Ceramic Society, OH, 1998.
[2]   Y. Sakabe, "Recent Development in Multilayer Ceramic Capacitors," pp. 3-15 in Multilayer Electronic Ceramic Devices. Ceramic Transactions, Vol. 97, Edited by J. H. Jean, T. K. Gupta, K. M. Nair and K. Niwa, American Ceramic Society, OH, 1998.

[3]   R. E. Mistler and E. R. Twiname, "Tape Casting Theory and Practice," The American Ceramic Society, Westerville, OH, 2000.

[4]   Kristoffersson, E. Roncari and C. Galassi, "Comparison of Different Binders for Water-Based Tape Casting of Alumina," *J. Eur. Ceram. Soc.*, **18**, 2123-31 (1998).

[5]   P. Nahass, W. E. Rhine, R. L. Pober, and H. K. Bowen, "A Comparison of Aqueous and Non-Aqueous Slurries for Tape-Casting, and Dimensional Stability in Green Tapes," pp. 355-64 in Materials and Processes in Microelectronic Systems. Ceramic Transactions, Vol. 15, Edited by K. M. Nair, R. Pohanka, and R. C. Buchanan, American Ceramic Society, OH, 1990.

[6]   H. W. Nesbitt, G. M. Bancroft, W. S. Fyfe, S. N. Karkhanis, A. Nishijima and S. Shin, "Thermodynamic Stability and Kinetics of Perovskite Dissolution," *Nature*, **289**[29], 358-62 (1981).

[7]   M. C. Blanco-Lopez, B. Rand and F. L. Riley, "The Properties of Aqueous Phase Suspensions of Barium Titanate," *J. Eur. Ceram. Soc.*, **17**, 281-87 (1997).

[8]   M. C. Blanco-Lopez, G. Fourlaris and F. L. Riley, "Interaction of Barium Titanate Powders with an Aqueous Suspending Medium," *J. Eur. Ceram. Soc.*, **18**, 2183-92 (1997).

[9]   Utech, "The Effect of Solution Chemistry on Barium Titanate Ceramics," M.S. Thesis in Solid State Science, The Pennsylvania State University, 1990.

[10] H. Yoon and B. I. Lee, "Barium Ion Leaching from Barium Titanate Powder in Water," *J. Mat. Sci.: Mat. Electro.*, **14**[3], 165-169 (2003).

[11] R. E. Chodelka, "The Aqueous Processing of Barium Titanate: Passivation, Dispersion, and Binder Formulations for Multilayer Capacitors," Ph. D. Dissertation in Materials Science and Engineering, University of Florida, 1996.

[12] X. Wang, B. I. Lee and L. Mann, "Dispersion of Barium Titanate with Polyaspartic Acid in Aqueous Media," *Colloids and Surfaces A*, **202**, 71-80 (2002).

[13] R. A. Kimel and J. H. Adair, "Aqueous Degradation and Chemical Passivation of Yttria-Tetragonally-Stabilized Zirconia at 25°C," *J. Am. Ceram. Soc.*, **85**[6], 1403-1408 (2002).

[14] Neubrand, R. Lindner and P. Hoffmann, "Room-Temperature Solubility Behavior of Barium Titanate in Aqueous Media," *J. Am. Ceram. Soc.*, **83**[4], 860-864 (2000).

[15] H. Yoon and B. I. Lee, "Processing of Barium Titanate Tapes with Different Binders for MLCC Applications – Part I: Optimization Using Design of Experiments," *J. Eur. Ceram. Soc.*, **24**[5], 739-752 (2004).

[16] H. Yoon and B. I. Lee, "Processing of Barium Titanate Tapes with Different Binders for MLCC Applications – Part II: Comparison of the Properties," *J. Eur. Ceram. Soc.*, **24**[5], 753-761 (2004).

[17] D. Christian, "Analytical Chemistry," Third Edition, John Wiley & Sons, New York, 1980.

[18] M. M. Lencka and R. E. Riman, "Thermodynamics of the Hydrothermal Synthesis of Calcium Titanate with Reference to Other Alkaline-Earth Titanates," *Chem. Mater.*, **7**, 18-25 (1995).

[19] Herard, A. Faivre and J. Lemaitre, "Surface Decontamination Treatments of Undoped $BaTiO_3$ – Part I: Powder and Green Body Properties," *J. Eur. Ceram. Soc.*, **15**, 135-44 (1995).

[20] K. Osseo-Asare, F. J. Arriagada and J. H. Adair, "Solubility Relationships in the Coprecipitation Synthesis of Barium Titanate: Heterogeneous Equilibria in the Ba-Ti-C_2O_4-H_2O System," pp. 47-53 in Ceramic Powder Science. Ceramic Transactions, Vol. 1, Edited by G. L. Messing, E. R. Fuller, Jr. and H. Hausner, American Ceramic Society, OH, 1988.

[21] S. W. Lu, B. I. Lee and L. A. Mann, "Characterization of carbonate on $BaTiO_3$ ceramic powders," *Mat. Res. Bull.*, **35**[8], 1303-12 (2000).

[22] Busca, V. Buscaglia, M. Leoni and P. Nanni, "Solid-State and Surface Spectroscopic Characterization of BaTiO3 Fine Powders," *Chem. Mater.*, **6**, 955-61 (1994).

[23] S. W. Lu, B. I. Lee and L. A. Mann, "Carbonation of Barium Titanate Powders Studied by FT-IR Technique," *Mat. Lett.*, **43**[3], 102-05 (2000).

[24] F. K. Hennings, C. Metzmacher and B. S. Schreinemacher, "Defect Chemistry and Microstructure of Hydrothermal Barium Titanate," *J. Am. Ceram. Soc.*, **84**[1], 179-82 (2001).

[25] R. Brezinski, "An Infrared Spectroscopy Atlas for the Coating Industry," Blue Bell, PA, 1991.

[26] C. Li and J. H. Jean, "Interaction between Dissolved Ba^{2+} and PAA-NH_4 Dispersant in Aqueous Barium Titanate Suspensions," *J. Am. Ceram. Soc.*, **85**[6], 1449-55 (2002).

[27] J. A. Lewis, "Binder Removal from Ceramics," *Annu. Rev. Mater. Sci.*, **27**, 147-73 (1997).

[28] Y. H. Hu, M. P. Harmer and D. M. Smyth, "Solubility of BaO in $BaTiO_3$," *J. Am. Ceram. Soc.*, **68**[9], 372-76 (1985).

[29] Kulcsar, "A Microstructure Study of Barium Titanate Ceramics," *J. Am. Ceram. Soc.*, **39**[1], 13-17 (1956).

[30] M. Levin, C. R. Robbins and H. F. MaMurdie, "Phase Diagrams for Ceramists," American Ceramic Society, Columbus, OH, 1964.

[31] K. Kinoshita and A. Yamaji, "Grain-Size Effects on Dielectric Properties in Barium Titanate Ceramics," *J. Appl. Phys.*, **47**[1], 371-73 (1976).

[32] Arlt, D. Hennings and G. D. With, "Dielectric Properties of Fine-Grained Barium Titanate Ceramics," *J. Appl. Phys.*, **58**[4], 1619-25 (1985).

[33] Herczog, "Application of Glass-Ceramics for Electronic Components and Circuits," *IEEE Trans., Parts, Hybrids, Packag.*, **PHP-9**[4], 247-56 (1973).

[34] J. Lee and K. Hong, "Roles of Ba/Ti Ratios in the Dielectric Properties of $BaTiO_3$ Ceramics," *J. Am. Ceram. Soc.*, **84**[9], 2001-06 (2001).

In: Trends in Materials Science Research
Editor: B.M. Caruta, pp. 219-244

ISBN: 1-59454-367-4
© 2006 Nova Science Publishers, Inc.

Chapter 9

SURFACE PROCESSES IN CHARGE DECAY OF ELECTRETS

G. A. Mekishev[1]

Department of Experimental Physics, University of Plovdiv 'P. Hilendarski',
Tzar Assen Str. 24, 4000 Plovdiv, Bulgaria

Abstract

The present paper is a review of the investigations devoted to the role of the sample surface in electret charge decay. Obtaining of electrets was carried out in a corona discharge. The set-up comprises trielectrode system – a corona electrode (needle), a grounded plate electrode and a grid (control electrode) in between. Metal mask with numerous apertures was used instead of the grid for obtaining electrets with island surface charge distribution. The electret surface potential was measured by the method of the vibrating electrode with compensation.

We studied the influence of the anisotropy, and pollution of quartz electrets on the surface potential decay. The behaviour of polymer electrets with island and uniform surface charge distributions was also studied. The results show that at low temperatures and high humidity the drift of the charges on the surface prevails over the transport processes through the bulk of the electret.

The results of studying the behaviour of polymer electrets obtained in corona discharge performed in different gas media or placed under condition of various pressures lower than atmospheric are reported. It was shown that the time dependence of the surface potential is described well by the differential equation for desorption and the dependence of the surface potential on pressure is satisfactorily described by an equation that is analogous to the Langmuir law of adsorption.

Next, the boundary surface potential of electrets according to Paschen's law was determined. An equation for the boundary surface potential was given and a "universal" curve was drawn. Experimental results show that if the surface potential was higher than the boundary surface potential, the charge decay was much more rapid.

Finally, the percolation model was used to analyse electret surface discharge of electrets stored at various conditions of controlled relative humidity for 250 days.

---

[1] E-mail address: mekishev@pu.acad.bg

All investigations performed show that under certain conditions surface processes should not be neglected and to extrapolate the results under conditions different from those of obtaining them is incorrect.

Introduction

An electret is a piece of dielectric or high resistivity material that produces an external electric field due to a preliminary polarization or electrification.

After preparing the first electrets the first theoretical understanding for the nature of the electret effect comes into being. Thanks to the efforts and contributions of many researches this understanding has ceaselessly been modified before the modern views for the processes responsible for the electret effect in dielectrics were achieved. Gross [1] has a great contribution to the electret theory progress and the development of the experimental technique.

At present there are two group theories for the explanation of the electret effect – the phenomenological theory and the injection theory.

The phenomenological models have arisen to explain mainly the properties and the behaviour of thermoelectrets. They can be classified in two basic groups.

The models in which the superposition principle is not used belong to the first group and it is assumed that the rate of decreasing is either proportional to the available instantaneous polarization or to the difference between the value of polarization in a given moment and the equilibrium value of polarization. This approach is advanced by Adams [2] and developed in the studies of Swann [3] and Gubkin [4].

The models where the superposition principle is suggested, i.e. the polarization is considered as a superposition of the individual responses to the individual electric fields, belong to the second group. This approach is developed in the studies of Gross [5], Wiseman and Feaster [6], Perlman and Meunier [7], Tilly [8].

Many authors have assumed that Ohm's law is valid and that injected charge decreases due to volume conductivity of the dielectrics under the action of the internal electric field.

According to a model proposed by Bogoroditzkiy et al [9] a main role in the forming of electret states in ceramic materials plays the injected charge.

Injection models describe the surface potential decay of electrets, high resistivity semiconductors and photoconductors as a result of injection of surface charge into the bulk, subsequent transport through the sample and a charge trapping and detrapping during transport [10-19].

However, in both cases – phenomenological and injection models – the processes in the bulk of the sample have been studied while the possible influence of the surface on charge relaxation has been neglected.

In [17] a surface conduction limited discharge has been pointed out as one of the various mechanisms, which may lead to the surface potential decay, but it is assumed that the surface conductivity is sufficiently small to neglect this component.

In [20] the charge decay in two samples of Teflon-FEP are compared, one of them being provided with a grounded metal strip on its charge surface. The decays are practically identical at a temperature of 200°C. The author concludes that the main self-drift of the charges is not along the surface. But at high temperatures, the transport processes through the bulk play a main role and the surface is no longer determinant.

The paper presents some results, which illustrate the role of the surface processes in surface potential decay of electrets.

The influence of the anisotropy, pollution and the surface relief of quartz electrets and the surface charge distribution of polymer electrets on the surface potential decay was studied. The behaviour of electrets obtained in corona discharge performed in different gas media or placed under conditions of various pressures lower than atmospheric was also studied.

An equation for the boundary surface potential of electrets according to Paschen`s law was given and a "universal" curve was drawn, and the percolation model was used to analyze electret surface discharge of electrets stored at various conditions of controlled relative humidity.

Experimental

Deposition of charges of positive or negative polarity onto the surface of the insulator is most suited for investigation of the surface processes. Convenient and wide applied method for deposition of charges onto the sample surface is the method of corona discharge [21 – 23].

The devices usually consist of a needle or a fine wire, called a corona electrode, and a plate electrode. In between them a third wire mesh electrode (a grid) is placed to control the process of charging of electrets. Near the corona electrode the air becomes ionized and the resulting ions are driven towards the plate electrode. The ionization is limited to a region close to the corona electrode and the drift region extends towards the plate electrode [22]. The advantages of corona charging are simplicity of the set-up required and speed of charging.

Fig.1 schematically shows the set-up used in our experiments. The set-up consists of a grounded plate electrode, a corona electrode (a needle) and a grid placed between them. The sample was put onto the grounded plate electrode. In some cases of nonmetallized samples some liquid, such as water, etc. could be put between it and the plate electrode, in order to improve the contact.

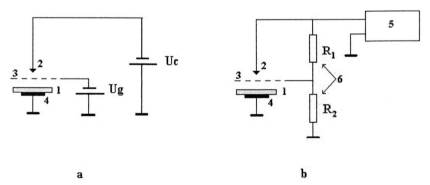

a) scheme with two independent voltage supply   b) scheme with one common voltage supply

Fig.1. Set-up for obtaining electrets

1 – sample ; 2 – corona electret; 3 – grid; 4 – plate electret; 5 – voltage supply; 6 – voltage divider; Uc – corona voltage supply ; Vg – grid voltage supply

In Fig. 1a a scheme with two independent voltage supplies is represented and in Fig.1b – with one common voltage supply. In the second case the grid was connected to a voltage divider.

The set-up of charging may be placed into a chamber connected to a vacuum pump and a gas inlet in order to perform charging in various gas media.

The surface potential of electrets was measured by the method of vibrating electrode with compensation [24]. The measurement device may be also located in a vacuum chamber connected with a gauge for measuring the pressure in the chamber if it needs to study the influence of under-pressure on surface potential decay.

Before charging the samples, quartz and polymers, were usually cleaned in an ultrasonic bath.

the eventual influence of humidity. The storage conditions and the initial values of the surface potential are presented in Table 2. The following designations have been used: T_s – temperature of storage, V_{eo} – initial values of the surface potential.

Table 2. Storage conditions and initial values of the surface potential for quartz electrets

Sample	T_s, $^\circ$C	Medium	V_{eo}, V
Y-cut	25	Argon	550
Z-cut	25	Argon	590
M	25	Argon	540

The dependence of the surface potential on the storage period in argon environment has been illustrated in fig.2. The same figure illustrates for comparison the curve $V_e \approx \exp(-t/\tau)$ referring in this case to bulk conductivity for a X-cut where $\tau = \varepsilon_0 \varepsilon \rho$ ($\varepsilon_0 = 8.85 \cdot 10^{-12}$ F/m, $\varepsilon = 4.5$, $\rho = 2 \cdot 10^{14}$ Ωm [27]).

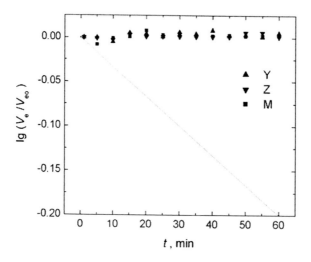

Fig.2. Dependence of surface potential of quartz electrets on time of storage in argon atmosphere :

Straight line is according to dependence $V_e \approx \exp(-t/\tau)$

The results obtained show sufficiently clearly that under the conditions specified above:

1. Conductivity anisotropy does not influence significantly the surface potential decay. The prolonged storage of the samples in an argon environment has demonstrated the same result. For example, for the samples of the Z-cut after 5 hours $V_e = 0.998 \ V_{eo}$, for the samples of molten quartz $V_e = 0.994 \ V_{eo}$, and for the Y-cut samples after a little more than 28 hours /1700 minutes/ $V_e = 0.976 \ V_{eo}$. We would like to note that the bulk conductivity for a Y-cut is 200 times less that than for a Z-cut.

2. Ohm's conductivity is not the reason for the surface potential decay of the quartz electrets. The dependence $V_e \approx \exp(-t/\tau)$ is much steeper than the experimental curves of the surface potential.

The experiment conducted and the results obtained allow making the assumes a significant role for the surface potential decay.

Influence of the Degree of Purity of the Surface

The further tests have the purpose to clarify the influence of the state of the surface and the degree of purity in particular on the surface potential decay. Tests were carried out with quartz samples, X-cut with thickness Le = 0,19mm, metalised in one side. Charging was carried out in a corona discharge at room ambient conditions and the corona voltage was $V_c = 5kV$. The results have been presented in fig.3.

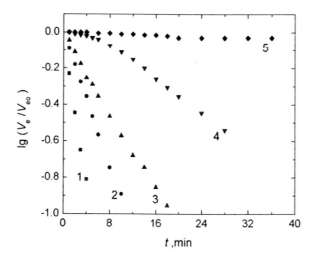

Fig.3. Influence of the degree of pollution surface potential decay. Degree of pollution decreases from 1 to 5 .

In fig.3 the curves from 1 to 5 correspond to different degrees of pollution of the samples decreasing from 1 to 5. Curve 4 was recorded after the samples were washed out before charging in an ultrasonic bath with spirit and distilled water and then dried out in the air at room ambience. Curve 5 refers to samples that were heated for a while at 100°C after the purification in the ultrasonic bath and were then charged and stored at 75°C. This temperature was selected in order not to activate the strongly bulk processes.

The results explicitly indicate that with the decrease in the degree of pollution (increase in the degree of purity) on the sample surface, the rate of the surface potential decay decreases. The decay rate is lowest in sample 5, which has the purest surface regardless of the fact that at this temperature there is an initial stimulation of the bulk processes.

According to [28] at a relative humidity of 70% the surface resistivity on polluted surface is $\rho_s = 2 \cdot 10^8 \, \Omega$ and on pure surface $\rho_s = 10^{13} \, \Omega$. This data allow for understanding the results presented in fig. 3.

The next experiment provides additional evidence for the availability of drift of charges on the sample surfaces.

Influence of the Mechanical Relief of the Surface

Samples of X-cut quartz with thickness $L = 0.53$mm have been studied. Two series were prepared with a different profile (relief) of the surface. A ditch was made on one of the surfaces of the samples of the first series, while the surface of the samples from the other series was left smooth. The form of the two series of samples is illustrated in fig.4.

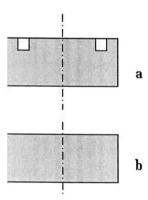

Fig.4. The form of X-cut quartz samples a) sample with ditch, b) sample without ditch

The samples were washed out in acetone in advance and then dried at room ambient temperature. Charging was carried out in corona discharge at room ambience. The corona electrode was fed with a negative voltage of –5kV. The samples from the two series were charged to approximately the same potential of the order of 1000V and were stored at room ambience [29].

For a period of 3 weeks the surface potential was measured. Its time dependence for the two series is presented in fig.5. After about 20 days the surface potential decayed to 0.7 from the initial value for the electrets with a ditch and to 0.2 from the initial value for the electrets without a ditch. These results show the greater stability of electrets with a ditch. They can be understood if we assume that at low temperatures and in the presence of humidity the discharge takes place mainly along the sample surface.

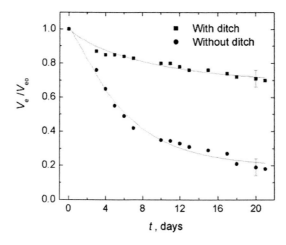

Fig.5. Effect of the X-cut sample surface relief on the surface potential decay.

Electrets with Islands Surface Charge Distribution

To create a definite mechanical relief of a thin polymer foil surface is practically impossible. Therefore a study of the electret behaviour in the case of a non-uniform charge distribution is needed, e.g., when the charge is deposited as separate islands, distributed over the electret surface. In [30] the behaviour of electrets with uniform and island surface charge distributions was compared.

The electrets with uniform charge distribution were obtained using a conventional three-electrode system (Fig.1). The electrets with island surface charge distribution were obtained using a modification of a set-up for obtaining electrets with uniform charge distribution. It consists of a grounded plate electrode, a metal mask with numerous apertures of a definite size and distribution of the apertures, and a corona electrode [30]. The metal mask functioned as a control electrode. The shape, the size and the distribution of the apertures in the mask determined the shape, the size and the distribution of the charged areas over the electret surface.

Samples of polytetrafluorethylene (PTFE) and polypropylene (PP) of the same thickness, 20μm, were investigated. The samples were pre-cleaned in an ultrasonic bath with acetone and rinsed successively with alcohol and distilled water and then dried. The samples were put on the grounded plate electrode. To obtain an electret with island charge distribution, the mask was put upon the samples. Voltage of –5kV was applied to the corona electrode, and a voltage of about –400V was applied to the mask or to the grid. The time of charging was 1 min. The electrets obtained with uniform and island charge distributions were stored in a desiccator under the same conditions: room temperature and relative humidity of 98%. The electret surface potential was measured out of the desiccator.

Pictures of uniform and island charge distributions are shown in Fig.6. Time dependences of the surface potential of PP electrets are shown in Fig.7. All points are mean values obtained from six samples. The results obtained for PTFE samples are analogous to those for the PP samples presented in Fig.7 [30]. As can be seen from the Figure, the surface potential decay for the electrets with island charge distribution (curve 1) is considerably slower than that for the electrets with uniform charge distribution (curve 2). The results of the experiment performed enable us to believe that at low temperatures and high humidity the drift of the charges along the surface prevails over the transport processes through the bulk of the electret.

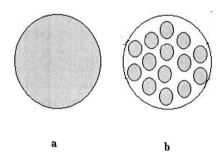

a   b

a) uniform charge distribution   b) island charge distribution

Fig.6. Pictures of charge distributions

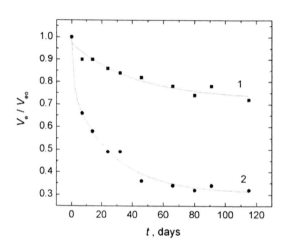

Fig.7. Time dependence of the surface potential of PP electrets obtained by negative corona ; 1 – island charge distribution; 2- uniform charge distribution

Furthermore, two series of samples of PP with island and uniform charge distribution were stored at 72°C, a temperature high enough to excite the bulk effects in PP. The results obtained are presented in Table 3 [30].

Table 3. Surface potential decay of PP electrets with a different charge distribution stored at 72°C [30].

Distribution	Surface potential		
	At $t=0$, (V_{eo}), V	After 2 hours at 72°C (V_e), V	$\dfrac{V_e}{V_{eo}}$
Uniform	364 ± 2	113 ± 7	$0 \cdot 31$
Island	307 ± 19	97 ± 7	$0 \cdot 31$

The results in Table 3 show that at high temperatures the electret surface is no longer determinant. Analogous results for the charge decay of two Teflon-FEP foils one of which has been provided with earthed metal strips on its charge side kept at 200°C for 5 min are reported in [20].

The results presented in Table 3 and those given in [20] show that at high temperatures the surface potential decay is due to various mechanisms, for example, charge injection from the surface and subsequent transport through the bulk, thermal bulk generation of charge carriers, drift of the free charges in the bulk to the surface, etc. All these mechanisms take place in the bulk of the samples and their propagation to the surface, which is at room temperature, is incorrect.

Electrets Obtained in Different Gas Media

The influence of gas medium in which the corona discharge was carried out on the properties of electrets has been studied by number of researches [31-36]. Different authors have investigated various aspects of behaviour of electrets obtained.

In [31] results of investigation of temperature dependence of surface potential of Teflon-FEP foils charged in different gases are reported. The electrets have been formed by negative spark discharges in the following gases: air, N_2, O_2, CO_2, SF_6, Ar. It was found that there is no significant difference in surface potential decay for electrets charging in different gases. On the basis of results obtained it was concluded that the surface potential decay is determined by the bulk properties of Teflon-FEP rather than the type of ions deposited onto the sample surface.

Raposo et al [32] have been performed thermally stimulated depolarization current of Teflon-FEP electrets. The samples have been charged in a corona discharge for two sets of charging times (10s and 30min) in various gases: N_2, O_2, CO_2, air and dry air. It was observed that the neutral activated species interact with polymer surface and create new surface traps and that heating charged-treated samples the created traps are eliminated.

Analogous measurements also of Teflon-FEP electrets obtained in corona discharge in various gases have been made in [33]. The results confirm the assumption advanced in [32]. It was supposed that the nature of the produced activated species depends on the atmosphere and that the neutral activated species modify the surface traps. This fact that the corona discharge modifies the sample surface is widely used for improvement of the adhesion ability of polymer [37, 38]. It was found that no injection of charge in the bulk of polymer takes place.

The time dependences of the surface potential of polymer electrets were studied for more than 200 days [35]. Samples of PTFE, PP (Bulgaria and Japan) and PET were charged in positive and negative corona in various gas media. The charging in a corona discharge was carried out in a special chamber in which the charging set-up was placed [35]. Depending on whether the gas heavier or lighter than air, it was run into the chamber from the bottom upwards or from the top downwards. The electrets obtained were kept on a metal pad under room conditions.

It was found that there is significant difference in surface potential decay for electrets charging in different gas atmospheres. For example, with PTFE electrets 240 day after they have been charged in negative corona $V_e \approx 0.9V_{eo}$ for the samples charged in air and $V_e = 0.6V_{eo}$ for those charged in N$_2$.

The experimental results are described well by the equation

$$V_e(t) = at^{-n}, \qquad n<1 \tag{1}$$

where a and n are constants.

In order to explain the results obtained on the basis of the injection model the following equation for the surface potential decay was derived:

$$V_e(t) = \frac{V_{eo}}{1+n}(\frac{t_{o\lambda}}{t})^n, t \geq t_{o\lambda} \tag{2}$$

where $t_{o\lambda}$ is the transit time of the leading front and V_{eo} is the initial surface potential. From the equation (2) and relation

$$t_{o\lambda} = \frac{L^2}{\mu V_{eo}} \tag{3}$$

where L is the sample thickness, the transit time for leading front $t_{o\lambda}$ and the charge carrier mobility μ were calculated. It should be noted that for the same material, depending on the gas used in corona discharge, different values for $t_{o\lambda}$ and μ are obtained. For example, with PTFE electrets $\mu = 1.13 \cdot 10^{-17} m^2/(V.s)$ for the samples charged in Ar atmosphere and $\mu = 9.10^{-18} m^2/(V.s)$ for O$_2$. It is likely that at room temperature the injection of charges in the bulk of the polymer does not determine the charge decay.

The effect of different gases used in obtaining electrets is different for the different materials and the different sign of the corona. Probably it is due to the different type of traps that are produced during the interaction of the gas molecules with the surface of the samples. It is also possible that the difference in the behaviour of the adsorbed gases is due to different electron structure of their molecules and the type of the interaction with surface. This conclusion is consistent with that reported in [32, 33].

The influence of corona polarity (positive or negative), on the electret surface potential thermograms and the kinetics of the charge decay in PTFE electrets, obtained in negative corona was studied in [39]. It was found that in the temperature dependence of the surface potential was observed a maximum, which is different for the electrets obtained in a positive and negative corona. In the case of negative corona the form and the magnitude of the peak change with the time of storage. It was shown [36] that the magnitude of the peak depends to a considerable extent on the gas used in the corona discharge. The presence of such a peak is supposed to be due to the availability of oxygen in gas medium, which results in the formation of dipoles with a certain dipole moment and as a consequence a dipole polarization was occurred. With increasing temperature, polarization decreases, while deposited charge remains almost constant, i.e. apparent increase of the surface potential is recorded [36].

An attempt to study by X-ray photoelectron spectroscopy the polymer electrets obtained by a corona discharge in air was made [40]. It was established a change in the binding energy for the C1s and O1s peaks and for the O1s/C1s ratio. These results confirm the assumption that the corona discharge creates additional traps at the surface layer of the samples.

In order to compare the time and temperature dependences of surface potential decay of electrets obtained in various gas atmospheres an additional experiment was carried out [41]. Nonmetalised samples of polypropylene with thickness of 20μm and diameter of 30 mm were studied. Before charging the samples was cleaned by alcohol in ultrasonic bath for 4 minutes, washed in distilled water and then dried.

The charging set-up (Fig.1b) was located into a vacuum chamber. The sample was placed onto the grounded electrode and was kept at at 20Pa for 24 hours. Then the chamber was again evacuated and filled up with the used gas until the pressure became equal to the atmospheric one. A constant voltage of –5kV was applied to the corona electrode for a period of 1 minute. The electrets obtained were stored in desiccator and the surface potential was measured out of the desiccator.

Because of the set-up used (Fig.1b) the initial potentials of electrets charging in different gas atmosphere are different [41]. This is due to the different critical corona voltage for the different gases used [34].

The time dependences of the surface potential for PP samples charged in different gas atmospheres (air, nitrogen and oxygen) are presented in Fig.8. It is seen that the decay rate is different for samples charged in different gases.

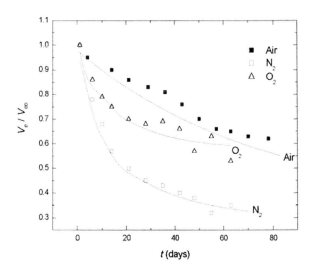

Fig.8. Time dependence of the surface potential of PP electrets obtained by negative corona in various gas media.

The temperature dependences of the surface potential decay also for PP samples charged in negative corona are presented in Fig.9. The heating rate for all the samples is the same-1.9K/min. For all the samples is observed no signification difference of surface potential decay.

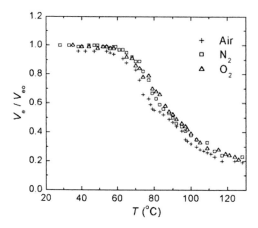

Fig.9. Temperature dependence of the surface potential of PP electrets obtained by negative corona in various gas media.

The results obtained are consistent with those presented in previous section. They enable us to conclude that at law temperatures the sample surface of high resistivity materials plays a dominant role. However at elevated temperatures the surface is no longer determinant and the bulk processes play the main role.

Electrets Stored at Low Pressure

The behaviour of electrets placed under conditions of various pressures lower than atmospheric has been studied by Wild and Stranathan [42]. The electrets have been made of carnauba wax and of polyvinyl acetate resin. The dissectible-capacitor method has been used for measuring the equivalent surface charge. It was supposed that the equivalent surface charge decreases because of the breakdown in the air gap during the measurement.

Recently, the results of investigations on the charge decay in polymer electrets stored at pressure lower than atmospheric was reported [43]. Now we shall set forth these results.

Non-metalised samples of polypropylene with thickness of 20μm were studied. All samples had a diameter of 30mm. Before charging the samples were cleaned in an ultrasonic bath with alcohol for 4 min, rinsed with distilled water and then dried under room conditions.

Charging of the samples was carried out by the corona discharge set-up. The voltage of the corona electrode was $U_c = -5$ kV and the charging time was 1min. The electret surface potential was limited by the grid potential and was of the order of -650 V.

The electret surface potential was measured by the method of vibrating electrode with compensation. The measuring device was located in a vacuum chamber connected to a gauge for measuring the pressure in the chamber.

After charging the samples were placed into the measuring device in the vacuum chamber, the initial surface potential was measured and then the chamber was evacuated and the surface potential was monitored over a period of 1h. The following pressures were called into being in the chamber: 20, 50, 100, 150, 200 and 300torr. The electrets stored at atmospheric pressure of 760 torr were also measured.

The surface potential versus storage time relationships for electrets stored at various pressures are presented in Fig.10. They show that surface potential decays rapidly during the first minute and after 10min. it stays practically constant. Furthermore, the lower the pressure in the chamber the higher the charge decays.

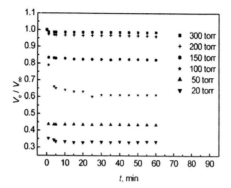

Fig.10. Time dependence of the surface potential of PP electrets at various pressure.

As the measurements of the surface potential at under-pressure were over, the electrets were removed and kept under room conditions, and the surface potential was measured periodically. Besides, it should be noted that electret surface potential was not changed after the samples had been removed from the vacuum chamber and placed under atmospheric pressure, i.e. a surface potential recovery is not observed. The dependence of the electret surface potential on the time of storage at room conditions after the samples had been taken out of the vacuum chamber were also studied. The results show [43, 44] that the lower the pressure in the vacuum chamber, the slower the surface potential decay after the electrets are removed from the chamber. The surface potential of electrets placed at $p = 300$torr decreases from $V_{eo} = 586$ V just after removal from the chamber to $V_e = 451$ V after 25 days under room conditions, i.e. $V_e/V_{eo} = 0.77$ and so their stability does not differ from that of the control samples ($p=760$ torr) for which the surface potential decreases from $V_{eo} = 695$V to $V_e=530$V for the same period of time, i.e. $V_e/V_{eo} = 0.76$.In the same time the surface potential of electrets, placed at $p=20$ torr decreases from $V_{eo} = 333$V just after removal from the chamber to $V_e = 325$V after 25 days under room conditions.

As it has already been mentioned, Wild and Stranathan explains the decrease of the equivalent surface charge with breaks in the air gap between the electret and the measurement electrode. This is a possible explanation. Another possible explanation is that under a pressure fall desorption of ions from the electret surface takes place, which leads to a decay of the surface potential.

In order to obtain additional information about the possible processes, an additional experiment was carried out. The fact we utilized was that the surface potential of electrets does not change when they are taken out of the vacuum chamber and placed under atmospheric pressure conditions.

A series of 8 samples was placed on a metal strip in the vacuum chamber after being charged in corona discharge and their initial surface potential was measured. They were left in the vacuum chamber for 30 minutes at 5 torr. After the samples were taken out of the vacuum chamber, their surface potential was measured again. The value obtained for the

V_e/V_{eo} ratio was of the order 0,33 [45]. This value did not differ significantly from the one for electrets positioned between two plate electrodes (fig.10). This allowed us at this stage to give our preference to the desorption processes from the electret surface.

The desorption equation in the case of monomolecular layer and the absence of interaction between the adsorbed ions and radicals can be written according to [46] as

$$\frac{d\theta}{dt} = -k_1\theta + k_2 p(1-\theta), \tag{4}$$

where $\theta = V_e/V_{eo}$ is assumed to be equal to the degree of surface level filling, k_1 and k_2 are rate constants of desorption and adsorption respectively.

At the initial moment $t = 0$, then $V_e = V_{eo}$, and thus $\theta = \theta_o = 1$.

The solution of the differential equation (4) at the condition pointed out is given by

$$\theta = a + b.e^{-ct}, \tag{5}$$

where

$$a = \frac{k_2 p}{k_1 + k_2 p}; \quad b = \frac{k_1}{k_1 + k_2 p}; \quad c = k_1 + k_2 p. \tag{6}$$

Fig.11 shows an example of best fits obtained by applying equation (5) to the experimental data presented in Fig.10 for three curves (20, 100 and 300 torr). The measured values are shown as points and the solid lines present theoretical results as given by equation (5). For all curves the coefficients of determination are within the range 0.90 - 0.97.

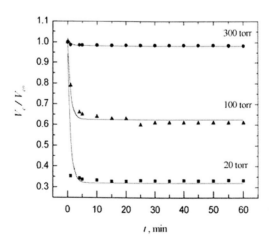

Fig.11. Time dependence of the surface potential of PP electrets . Points are the measured values, solid lines present theoretical results as given by equation (5).

● - 300 torr; ▲- 100 torr ; ■ – 20 torr.

From the equation (6) follows that:

$$a + b = 1 \qquad (7)$$

Values of the coefficients a, b and c calculated from experimental data, by means of fitting the data of fig.10 to the equation (5) and the sum of a and b are presented in [44] and show a good agreement of the theoretical model with the experimental data.

From the data presented in Fig.10 the dependence of the surface potential on pressure is obtained. The results obtained are shown in Fig.12. The theoretical curve is also drawn and is described by equation

$$\theta = \frac{p}{(A + Bp)} \ , \qquad (8)$$

which is analogous to the Langmuir law of adsorption.

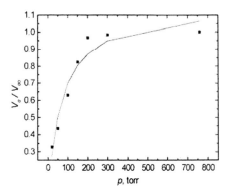

Fig.12. Dependence of the surface potential of PP electrets on pressure.

It should be noted that the constants k_1 and k_2 in this case are functions of the degree of surface level filling.

Surface Potential Decay and Paschen's Low

Many papers point out that in a number of techniques for obtaining electrets and measuring their surface charge, and also when using them in various construction, it should be taken into account that the maximum surface charge density is limited by the electric strength of the ambient air [34, 47-51]. The ambient electric strength depends on the geometry of the electrode-electret gap, on the gas composition and its pressure in this gap according to Paschen's law. A breakdown will occur when the voltage between the nonmetallized surface of the electret and the near electrode exceeds the Paschen's curve. As a result of the breakdown, the electret charge is reduced to a certain value at which the voltage curve passes under the Paschen's curve [50].

Usually, all the researchers discuss σ_{max} dependences of electret thickness L_e at a fixed value of the dielectric constant ε_e. Besides, the dependence $\sigma_{max} = f(L_e)$ is given with numerical values for the coefficients that, in fact, are different for the various parts of Paschen's curve. In all cases it is considered that the effect of Paschen's law upon σ_{max}, is mostly related to an electret that is between two electrodes and is not related to one kept in a free (with a gap between non-metalised surface and the near electrode, tending to infinity).

In [52] an equation for the surface potential of the electret, of more general kind with possibility of using different parts of Paschen's curve and analyzing the equation obtained, for a case of free electret storage is represented.

In our discussions we will follow Rose [50]. Usually, the construction of the electret transducers and schemes for measuring of the surface potential V_e contains an electret and two electrodes whereupon one of them is at a certain distance from the electret surface. The voltage between non-metalised surface of the electret and the upper electrode in the air gap is:

$$U = \frac{\varepsilon_e d}{L_e + \varepsilon_e d} V_e \tag{9}$$

where L_e is the thickness of the electret , ε_e is the dielectric constant of the electret material, d – the size of the electret-electrode gap, and V_e - the electret surface potential.

It is assumed, in formula (9), that dielectric constant of the air $\varepsilon = 1$.

To determine when a discharge is going to occur we have only to calculate the voltage in the air gap between the non-metalised surface of electret and upper plate electrode as a function of its size d by formula (9) and to plot the data obtained on a diagram with Paschen's curve for air plotted on, as well [47]. Changing V_e, we can determine that value of U and its corresponding value of the electret surface potential V_{eb}, at which a discharge is possible between the electret surface and the electrode according to Paschen's law [50, 51].

The boundary value V_{eb} at atmospheric pressure could be determined by the condition that a breakdown occurs when the air gap voltage curve $U = U(d)$ crosses Paschen's curve or is tangent to it. The part of Paschen's curve corresponding to real values (for the electrets) of the air gap d, may be approximated with the equation:

$$U_p = a + bd \tag{10}$$

In the intersection points $U_P = U$. Taking into account (9) and (10) we obtain the equation

$$\varepsilon_e b d^2 + (a\varepsilon_e + bL_e - \varepsilon_e V_e)d + aL_e = 0 \tag{11}$$

The equation (11) will have a solution if the following condition is satisfied

$$D \equiv (a\varepsilon_e + bL_e - \varepsilon_e V_e)^2 - 4ab\varepsilon_e L_e \geq 0 \qquad (12)$$

The case when $D = 0$ corresponds to an osculation of the two curves. On the condition $D = 0$ the surface potential V_e is equal to the boundary surface potential V_{eb} and for physical reasons it follows that the boundary surface potential of the electret is given by

$$V_{eb} = a + b\frac{L_e}{\varepsilon_e} + 2\sqrt{ab}\sqrt{\frac{L_e}{\varepsilon_e}} \qquad (13)$$

From equation (11) on the condition $D = 0$ and taking into account (13) we can determine the size of the air gap d, at which an osculation of the two curves occur at a given ratio L_e/ε_e

$$d = \sqrt{\frac{a}{b}}\sqrt{\frac{L_e}{\varepsilon_e}} \qquad (14)$$

The formula obtained (13) is analogous by form to the corresponding formulas reported in [49, 50]. But in contrast to the formulas in [49, 50], the formula (13) has a greater generality.

First, from equation (13) one can see that V_{eb} depends on L_e and ε_e through their ratio (L_e/ε_e). We will call this ratio a normalized thickness of the electret. However much the thickness L_e and the dielectric constant ε_e of the two electrets differ, if their normalized thicknesses L_e/ε_e are the same, their boundary surface potentials V_{eb} are also equal, i.e. to identical normalized thickness, corresponds an identical boundary surface potential. It enables us to plot a "universal" curve $V_{eb} = V_{eb}(L_e/\varepsilon_e)$, corresponding to the air pressure of 760 torr. Such a "universal" curve, corresponding to a pressure of 760 torr, is built up in Fig.13. As the normalized thickness of the electret is determined the boundary surface potential of the electret can be immediately determined by this curve.

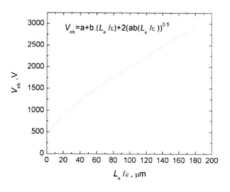

Fig.13. An "universal" curve build up according to equation (12).

Second, to build up a curve of the above mentioned kind, or to calculate the boundary surface potential V_{eb} by formula (13) we should know the values of the parameters a and b. Different authors give different values of a and b [50, 53-55]. Moreover, as it is pointed out in [53], the slope of Paschen's curve is getting smaller for gaps larger than 100μm. Therefore, before determining the values of the parameters a and b, we should determine that size of the air gap at which an osculation will occur of the curve $U = U(d)$ with Paschen's curve. The results of the calculations by formula (14) with parameters $a = 300$ V and $b = 7.10^6$ V/m, are given in Table 4.

Table 4. A size of the air gap at which an osculation will occur of the curve $U = U(d)$ with Paschen curve.

L_e / ε_e	1	5	10	20	40	60	80	100	150	200	225
d, μm	6.25	14,7	20.7	29.3	41.4	50.8	58.8	65.5	80.2	92,6	98.2

In case of polymer foil electrets the air gap $d < 100$ μm because of the small thickness of the electrets. With quartz electrets whose thickness is $L_e \leq 1$ mm and $\varepsilon_e = 4.5$, $(L_e / \varepsilon_e) \leq 222$ μm and thus $d < 100$ μm. With ceramic electrets the dielectric constant ε_e is much higher and therefore $d < 100$ μm even for samples considerably thicker than 1 mm. Hence, in a great number of practical cases, we can use the part of Paschen's curve relevant to air gaps that are smaller than 100 μm. Shaffert [53] gives a value of the parameters $a = 300$ V and $b = 6.2 \cdot 10^6$ V/m. Scanavy [54] gives a value of the parameter $b = 7.10^6$ V/m. These values are different from the ones Rose [50] and Lushteikin [49] have assumed: $b = 4.10^6$ V/m and $b = 4.10^4$ V/m correspondingly. As the experiment with PTFE electrets has shown, the best coincidence with experimental data is obtained if $b = 7.10^6$ V/m. That is why we have picked up $a = 300$ V and $b = 7.10^6$ V/m, and with these values we plot the "universal" curve , shown in Fig.13.

Third, the analysis of formula (13) shows that the boundary surface potential V_{eb} only depends on the normalized thickness of the electret and does not depend on the conditions the electret is under. Therefore, it should be expected that when the electret is not located between two electrodes and is kept in a free state, the surface potential decay will be much more rapid if the surface potential is higher than the boundary surface potential of the electret.

To verify the corollary, which was made above, a time dependence of electrets of 20μm thick PET was measured. Non-metalised samples were used and were charged in a corona discharge, up to a preliminary set up surface potential. Two series of samples were studied – one of them had a surface potential V_e, higher than the boundary surface potential, i.e. $V_e > V_{eb}$, and the other – had a surface potential lower than the boundary surface potential, i.e. $V_e < V_{eb}$. The samples were kept in open state on a metal plate placed in a desiccator, under conditions of high humidity to accelerate the charge decay. The surface potential was measured out of the desiccator. The results obtained are presented in Fig.14. They show that if the surface potential is higher than the boundary surface potential, the charge decay would be much more rapid compared to the case when the surface potential is lower than the boundary surface potential.

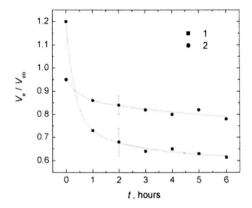

Fig.14. Time depending of PTFE electrets stored at high humidity (98 % RH)
1-surface potential is higher than boundary surface potential
2-surface potential is lower than boundary surface potential

Results presented in this section show that Paschen`s law holds also for the cases in which the charged surface is far from any metals or dielectrics. It is likely that there is an interrelation between Paschen`s law and the desorption prosseses from charged surfaces. But there is no evidence, which proves this hypothesis.

Perfolation Model of Electrets

The behaviour of electrets stored after under conditions of various humidity has been studied by many authors [42, 56, 57].

Wild and Stranathan [42] were found that the recovery of charge shown by electrets which were dried after being kept in high humidity for long periods may be due to further decay of the heterocharge.

Ribeiro et al were studied β-PVDF polarized in humid air (50%RH). They discard the hypothesis of surface conductivity and they consider that an increase in the sample conductivity is due to the water in the bulk.

The results of investigation of polymer electrets subjected to a corona discharge and stored at various humidity values are reported and the percolation model is used to explain the behaviour of electrets [58,59].

Non-metalised samples of polypropylene have been studied. The sample thickness was 20μm and all samples had a diameter 30mm. Samples were cleaned for 4 min in ultrasonic bath with alcohol, rinsed with distilled water and dried under room conditions. Before charging in a corona discharge, they have been stored in desiccators at various relative humidity (see table 5) for 50 days at room temperature.

Table 5. Relative humidity (RH) in the different desiccators.

No of desiccators	1	2	3	4	5
Relative humidity, %	0	32-33	52-55	75-76	98-100

Just before being put into the corona discharge, each of the samples was placed on a metal electrode, all of them 30 mm in diameter. The voltage of corona electrode was −5kV and the charging time was 1min. The electret surface potential (V_{eo}) was limited by the grid potential and was of the order of −670 V.

After charging the samples, the initial surface potential (V_{eo}) was measured. Then the electrets were put in desiccators at room temperature and the surface potential (V_e) was measured periodically out of the desiccators.

A few groups of samples have been studied [58, 59]: the first one comprising the numbers 51, 52,...55; the second one with designations 15, 25,...55 and the third one designated as 11, 22,...55. The first figure stands for the desiccator where the samples were placed before charging in the corona discharge; the second one indicates the desiccator where they were placed after charging. We will report the results of the third group: 11, 22,...55. Other results can be found in [58, 59].

Fig.15 represent the time dependence of the surface potential of the third group. All points are mean values obtained from six samples.

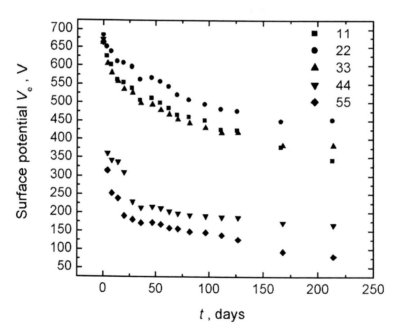

Fig.15. Time dependence of the surface potential of PP electrets, stored at various humidity conditions.

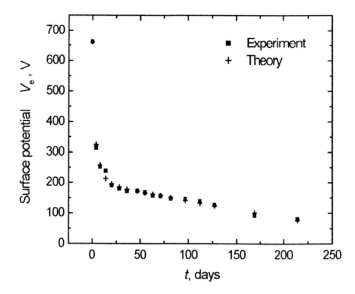

Fig.16. Comparison between two experimental data and theoretical calculations .

The results presented in Fig.15 show

- first, no matter if the samples before charging are kept at various relative humidities their behaviour is determined mainly by the storage conditions after charging;
- second, considerable charge decay is observed if the electrets are kept after charging at relative humidity higher than 55%;
- third, the difference in charge decay of electrets from different series should not be explained as a result of injection of the surface charge into the bulk and its subsequent transport through the sample.

If an electret is stored at certain humidity adsorption of water molecules from the environment take place on the surface and water-molecule clusters are formed. The aggregation of these clusters generates a two-dimensional (2D) percolation lattice. The concentration of the adsorbed molecules varies as a function of time, i.e. the structure and density of the percolation lattice also change with time. A model that account for these changes is the model developed by Kuzmin and Tairov [60-62].

The percolation model proposed by Kuzmin and Tairov [60-62] describes charge transport resulting from electret surface discharge caused by environmental humidity. This model is based on the simultaneous utilization of the percolation theory and the Kolmogorov's concept for a two-dimensional (2D) kinetics of nucleus formation.

As the concentration of the adsorbed water molecules gets higher, the individual clusters may interconnect and hence create an unbound cluster. The unbound cluster is formed when the fraction of the surface covered by adsorbed molecules θ reaches the percolation threshold or the critical percolation density θ_c, i.e. when $\theta = \theta_c$. According to [63] in the case of a 2D-percolation lattice $\theta_c = 0.44 \pm 0.02$. Since water conductivity is much greater than that for the material of the electret itself, the unbound cluster shunts the surface of the electret and causes fast neutralisation of the electret charge and surface potential decay is observed. Consequently, the fraction from the electret surface that interconnected with the unbound cluster does not contribute to the value of the surface potential. So the time dependence of the surface potential may be described by the following equation:

$$V_e(t) = V_{eo}[1 - P(\theta(t))],$$

(15)

where V_{eo} is the surface potential of the electret, measured just after the electret was obtained, and $P(\theta)$ is the density of the percolation cluster. $P(\theta)$ defines the probability for a randomly chosen point from the surface to belong to the specified cluster, and it represents a dimensionless quantity. The density of the percolation cluster is a function of the surface fraction $\theta(t)$ covered by the adsorbed molecules at a moment t.

In order to define $\theta(t)$ Kuzmin and Tairov assume the following [60, 62]: a) the mean area of the incipient adsorption centres called nuclei is infinitely small in comparison with the total surface of the electret; b) all the adsorption centres are generated randomly on the surface of the electret at a finite rate $\alpha(t)$ per unit area; c) all the adsorption centres have the same convex shape; d) as soon as they are generated and adsorb water molecules all the nuclei have the same rate of the growth $v(t)$.

At the moment of charging $t = \tau_0$ the following dependence describes $\theta(t)$:

$$\theta(t) = 1 - q(\tau_0, t)\exp\left[-\int_{\tau_0}^{t} \alpha(\xi) S(R(\xi, t)) d\xi \right]$$

(16)

where $q(\tau_0, t)$ is the fraction of the surface that has remained free from the adsorbed phase at the moment of charging τ, $R(\xi, t)$ is the radius of those nuclei which have originated after the sample was charged, and $S(R)$ is the surface of nuclei having radius R.

Provided that before charging the samples have been stored in a controlled stationary environment a thermodynamic equilibrium is established. Then, if we assume that the samples are charged at a moment $\tau_0 = 0$, the function $q(\tau_0, t)$ can be replaced by $q(0)$.

Taking that into account and writing $S(R) = \pi\left(\int_{t'}^{t} v(\xi) d\xi\right)^2$, the formula (16) may be represented as:

$$\theta(t) = 1 - q(0)\exp\left[-\int_{0}^{t} \alpha(t')\pi\left(\int_{t'}^{t} v(\xi) d\xi\right)^2 dt' \right]$$

(17)

where $\alpha(t)$ is the rate of nuclei formation at the moment t', and $v(\xi)$ is the rate of nuclei growth as a result of water molecule adsorption. The authors propose the following law for $\alpha(t)$:

$$a(t) = \beta\delta(t) + a_o \qquad (18)$$

where β is the concentration of the nuclei generated at the time of charging, $\delta(t)$ is the Dirac pulse function, α_0 is the steady rate of nuclei generation per unit area after the electrets have been obtained. The rate of nucleus growth $v(t)$ is supposed to go down by an exponential law:

$$v(t) = v_{01}\exp(-t/\tau) + v_{02} \qquad (19)$$

where v_{01} and v_{02} are constants.

Based on the regression analysis the authors suggest the following dependencies for density of the percolation cluster [62]:

$$0 \le \theta < \theta_c \qquad P(\theta) = 0$$

$$\theta_c < \theta \le 0.533 \qquad P(\theta) = \sum_{i=0}^{4} a_i (\theta - \theta_c)^i \qquad (20)$$

$$0.533 < \theta \le 1 \qquad P(\theta) = \theta$$

where $\theta_c = 0.428$, and this value is close to the one reported in [63] for the 2D-percolation lattice. The values of coefficients a_i are following : a_o = -1.042·10^{-2} ; a_1 = -21.26; a_2 = -3.959·10^2; a_3 = 3.549·10^3; a_4 = -1.178·10^6.

It is shown in [61] that according to the moment when an unbounded cluster appears there are three different types of transitions: a) the unbounded cluster is formed after some time since charging the electret; b) the unbounded cluster originates at the moment the charging is accomplished; and c) the unbounded cluster has already existed at the time of charging.

On the basis of the percolation model the values of the parameters $q(0)$, α_o, v_{o1}, and v_{o1} were calculated [59]. The time dependence of surface potential for sample 55 obtained from experimental data and the theoretical calculation are presented in Fig.16. The results obtained show a good conformity of theory and experiment.

The good agreement between the experimental and theoretical curves shows that the percolation model describes well the electret behaviour under various humidity conditions. The values of q are the same for all the samples ($q = 0.57$), which is in good agreement with the theoretical estimate $(q \approx 0.5715)$ given for this case in [61]. According to the estimate given in [62], a percolation transition may occur if the relative humidity is higher than 60%. The results we obtained did not confirm this, and showed that such a transition may also take place at lower relative humidities.

One should assume that moisture adsorption is stimulated by electrization, and further cluster growth depends on the storage medium after the charging of electrets. After the unbounded cluster is formed, when $\theta = \theta_c$ the further growth of $P(\theta)$ is due to the fact that already existing isolated clusters or newly formed clusters join it. The clusters that remain unconnected to the unbounded cluster are ineffective with reference to electrical conductivity and do not contribute to the surface potential decay. But, the so-called thickenings, duplication chains or dead ends will contribute to the surface discharge of the electret.

According to [59] for humidity values as high as 55% v_{01} does not practically change and v_{02} grows almost insignificantly. When the humidity goes over 55%, v_{01} and v_{02} increase rapidly, while α does not change. Therefore, the rate of nuclei growth is decisive for the unbounded cluster kinetics at these humidity values.

The model of Kuzmin and Tairov is the first one, which is based on the dynamic percolation theory. The model shows a good agreement with the experimental data. But some aspects of the model are needed to be précised.

Conclusion

The results presented and discussed in the paper make it clear that the role of surface potential decay of electrets should be taken into account. The results don't discard in any way available injection and other models. If the temperature is high enough or charge decay in photoconductors is studied injection models enable one to explain the experimental results well. Our investigation only show that it could not be consider that the high temperature only accelerate the processes in the electrets, which at room temperature take place slowly. In the different temperature ranges the different mechanisms could be manifested themselves and it is incorrect to extrapolate the results under conditions different from those of obtaining them.

References

[1]   Gross B. *Proc. 5th Intern. Symp. Electrets* (ISE 5); Heidelberg, (IEEE, NY), 1985 pp 9-46
[2]   Adams E. P. *J. Franklin Inst.* 1927, 204, 469-486.
[3]   Swan W. F. G. *J. Franklin Inst.* 1953, 255, 513-530.
[4]   Gubkin A. N. *Zh. Techn. Fiz.* 1957, 27, 1954-1968 (in Russian)
[5]   Gross. B. *J. Electrochem. Soc.* 1968, 115, 376-381.
[6]   Wiseman G.; Feaster G. *J. Chem. Phys.* 1957, 26, 521-527.
[7]   Perlman M. M.; Meunier J. L. *J. Appl. Phys.* 1965, 36, 420-427.
[8]   Tilley D. E. *J. Appl. Phys.* 1967, 38, 2543-2546.
[9]   Bogoroditzkiy N. P.; Tairova D. A.; Sorokin V. S. *Fiz. Tverdogo Tela* (Solid State Physics) 1964, 6, 2301-2306. (in Russian).
[10] Wintle H. J. *J. Appl. Phys.* 1970, 41, 4004-4007.
[11] Batra J. P.; Kanazawa K. K. *J. Appl. Phys.* 1971, 42, 1124-1130.
[12] Kanazawa K. K.; Batra J. P.; Wintle H. J. *J. Appl. Phys.* 1972, 43, 719-720.
[13] Kanazawa K. K.; Batra J. P. *J. Appl. Phys.* 1972, 43, 1845-1853.

[14] Rudenko A. I.; Arkhipov V. I. *J. Electrostatics*. 1978, 4, 309-323.

[15] Moreno R. A.; Figueiredo M. T. *Proc. 5th Intern. Symp. Electrets (ISE 5);* Heidelberg,(IEEE,NY), 1985, pp 283-286

[16] Berlepsch H. von J. Phys. D: *Appl. Phys.* 1985, 18, 1155-1170.

[17] Kasap S. O. J. *Electrostatics*. 1989, 22, 69-90.

[18] Kasap S. O.; Bhattacharyya A.; Liang Z. *J. Appl. Phys.* 1992, 31, 72-80.

[19] Amjadi H. *IEEE Trans. Dielectr. Electr. Insulation.* 2000, 7, 222-228.

[20] Turnhout J. van. in Electrets: Sessler G. M.: Ed; *Topics in Applied Physics*,Vol.33; Springer Verlag, Berlin-Haidelberg-New York, 1980, pp 180-181

[21] Sessler G. M. in Electrets: Sessler G. M.: Ed.; *Topics in Applied Physics*, Vol.33, Springer Verlag, Berlin-Haidelberg-New York, 1980, pp 30-31

[22] Giacometti J. A.; Oliveira O. N. Jr. *IEEE Trans. Dielectr. Electr. Insulation.* 1992, 27, 924-943.

[23] Leal Ferreira G. F.; Figueiredo M. T. *IEEE Trans. Dielectr. Electr. Insulation.* 1992, 27, 719-738.

[24] Reedyk C.; Perlman M. M. *J. Electrochem. Soc.* 1968, 115, 49-51.

[25] Bogoroditzkiy N. P.; Passinkov V. V.; Tareev B. M. *Electrotechnical materials. Energia*, Leningrad. 1977, p 44. (in Russian)

[26] Mekishev G. A.; Tairov V. N. *Dep. Viniti.* 9.07.1981 № 3394-81 Dep. (in Russian)

[27] Passinkov N. V. *Materialy elektronnoy techniki.* Visschaya shkola, Moskow. 1980.

[28] Bogoroditzkiy N. P.; Passinkov V. V.; Tareev B. M. *Electrotechnical materials. Energia*, Leningrad. 1977, p 48. (in Russian)

[29] Mekishev G. A.; Prikhodchenko V. A.; Tairov V. N; Karmazova P. G. *Bulg. J. Phys.* 1986, 13, 58-64.

[30] Karmazova P. G.; Mekishev G. A. *Europhys. Lett.* 1992, 19, 481-484.

[31] Turnhout J. van. in Electrets: Sessler G. M.: Ed.; *Topics in Applied Physics*, Vol.33, Springer Verlag, Berlin-Haidelberg-New York, 1980, p181

[32] Raposo M.; Ribeiro P. A.; Giacometti J. A.;Bento M. Amalia; Marat-Mendes J. N. *Proc.7th Intern. Symp. Electrets (ISE 7),*Berlin (IEEE,NY),1991, pp 687-692.

[33] Marat-Mendes J. N.; Raposo M.; Ribeiro P. A. in *Equilibrium Structure and Properties of Surface and Interfaces*, Goins A. ; Stocks G. M. Eds.; Plenum Pres, New York, 1992, 353-359.

[34] Paajanen M.; Wegener M.; Gerhard-Multhaupt R. *J Phys. D: Appl. Phys.* 2001, 34, 2482-2488.

[35] Karmazova P. G.; Mintchev M. S.; Mekishev G. A. *Bulg. J. Phys.* 1993, 20, 42-51.

[36] Karmazova P. G.; Mekishev G. A. *Bulg. J. Phys.* 1990, 17, 454-458.

[37] Stradal M.; Goring D. A. I. Canadian *J. Chem. Engineering.* 1975, 53, 427-430.

[38] Kusabiraki M. *Jpn J. Appl. Physics.* 1990, 29, 2809-2814.

[39] Karmazova P. G.; Mekishev G. A. *Sci. Works Plovdiv University.* 1984, 22, 17-21. (in Bulgarian)

[40] Karmazova P. G.; Marinova T. S.; Krastev V. I. Mekishev G. A. *Bulg. Chem.. communications.* 1995, 28, 94-104.

[41] Yovcheva T.; Mekishev G.; Nedev S. *Journal of Optoelectronics and Advanced Materials.* 2005,7, 234-240.

[42] Wild J. W.; Stranathan J. D. *J Chem Phys.* 1957, 27, 1055-1059.

[43] Mekishev G. A.; Yovcheva T. A.; Guentcheva E.; Nedev S. *J. Mater. Sci.: Materials in Electronics.* 2003, 14, 779-780.

[44] Mekishev G. A.; Yovcheva T. A.; Guentcheva E. Submited to *J. Electrostatics.*

[45] Yovcheva T.A.; Mekishev G.A.Unpublished data.

[46] Morrison S. The Chemical Physics of Surfaces, Mir, Moscow. 1980, p286./ in Russian /.

[47] Sessler G. M.; in Electrets; Sessler G. M.; Ed.; *Topics in Applied Physics*, Vol.33, Springer Verlag, Berlin-Haidelberg-New York, 1980, pp 21-22.

[48] Gubkin A. N. *Electrets, Nauka*, Moskow. 1978, (in Russian)

[49] Lushteykin G. A. *Polymer electrets, Chimia*, Moscow, 1976, (in Russian)

[50] Roos J. *J. Appl. Phys.* 1969, 40, 3135-3139.

[51] Palaia F. L. Jr; Catlin A. *J. Chem. Phys.* 1970, 52, 3651-3656.

[52] Mekishev G. A. *Investigation of the influence of free charges on the electret effect in dielectrics.* Ph. D Disertation, LETI, Leningrad, 1981./ in Russian /

[53] Shaffert R. *Electrophotography, Mir*, Moscow. 1968, (in Russian)

[54] Scanavy G. I. *Physics of dielectrics*, GIPML, Moscow, 1958, (in Russian)

[55] Knoll M.; Ollendorff F.; Rampe R., *Gaseentladungstabellen*, Springer-Verlag, Berlin, 1939, p.83.

[56] Amjadi H. *TEEE Trans. Dielectrics Electr. Insulation.* 1999, 6, 236-241.

[57] Ribeiro P. A.; Giacometti J. A.; Raposo M.; Marat-Mendes *J. N . Proc.7th Intern. Symp. Electrets (ISE 7),*Berlin (IEEE,NY),1991, pp322-327.

[58] Mekishev G. A.; Yovcheva T. A.; Karamazova P. G.; Marinov A. *T. in Future directions in thin film science and technology* Marshall J. M.; Kirov N.;Vavrek A.; Maud J. M. Ed.; 1996,pp.441-444

[59] Yovcheva T. A.; Mekishev G. A.; Marinov A. T. *J. Phys.: Condens. Matter.* 2004, 16, 455-464.

[60] Kuzmin Y. I. *Proc. 10th Intern. Symp. Electrets (ISE 10)* Athens, (IEEE,NY) 1999, pp 55-58.

[61] Kuzmin Y. I.;Tairov V. N. Zh. *Techn. Fiz.* 1984, 54, 964-965. (in Russian)

[62] Scher H.; Zallen R. *J. Chem. Phys.* 1970, 45, 3759-3761.

[63] Scher H.; Zallen R. *J. Chem. Phys.* 1970, 45, 3759-3761.

In: Trends in Materials Science Research
Editor: B.M. Caruta, pp. 245-260

ISBN 1-59454-367-4
© 2006 Nova Science Publishers, Inc.

Chapter 10

THEORETICAL STUDIES OF Q1D ORGANIC CONDUCTORS: A PERSONAL REVIEW

Vladan Čelebonović[*]
Institute of Physics, Pregrevica 118,
11080 Zemun-Beograd, Serbia and Montenegro

Abstract

The aim of this contribution is to review some aspects of theoretical studies of a family of quasi one-dimensional (Q1D) organic conductors known as the *Bechgaard salts*. In order to make it personal, it will retrace the evolution of the author's interest in this field.

The generalized aim was the calculation of the electrical conductivity of the Bechgaard salts and gaining some knowledge on the equation of state and thermal properties of these materials. The review ends with some ideas about the possible future developement of the field.

Keywords: Bechgaard salts; Electrical conductivity; Memory function method; Hubbard model

1 Introduction

I have entered the Institute of Physics (IoP) in Beograd in the spring of 1985. Three years before that I have graduated in physics at the University of Beograd, and after some time spent in looking (in vain) for a job in various research establishements was delighted by the fact that I was finally accepted in IoP. I have entered the Laboratory for interdisciplinary studies, and my research field was broadly defined as " the study of the behaviour of materials under high pressure" . My immediate task was to calculate the electrical conductivity

---

[*]E-mail address: vladan@phy.bg.ac.yu

under high static pressure of several *normal* metals. If I suceed in the calculation, my job in IoP would became stable. If not.... My first contact with the organic conductors dates from that hectic period, when I discovered the existence of these curious materials in the literature.

My real encounter with the organic conductors occured in Paris in France a couple of years later. In the spring of 1987., I was awarded a French governement schollarship for a visit of 8 months to Laboratoire des interactions moleculaires et hautes pressions (LIMHP) at Université Paris-Nord. The aim of my visit was to learn experimental techniques of high pressure research.

While working in LIMHP, I managed (through a friend of a friend from Belgrade) to arrange a visit to Laboratoire de Physique des Solides du CNRS on the campus of Université Paris *XI* in Orsay. I was to be received by "someone" named Denis Jérome. Even while I am writing this, after nearly 17 years, I am smiling when I think that on the day when my visit was being fixed, this name did not mean anything special to me! Luckily, the afternoon before I was due to go to Orsay, I received a reprint of a paper from "Contemporary Physics" from which I learnt that the man whom I was going to meet actually discovered most of the facts known about quasi one-dimensional (Q1D) organic conductors.

The rest of the story is "condensed history". The following day I went to Orsay and met D.Jérome. He received me extremely kindly, we had a friendly conversation, focused on high pressure and organic conductors, and somewhere around noon we went to the canteen. While there, I started thinking that it would be "great" if I could somehow get the possibility to work for my Ph.D. in Jérome's group. However, I was in a dilemma how to ask him such a thing after meeting him for the first time in my life. Whether or not he felt my thoughts I do not know. The important thing happened when in the course of the lunch we came to delicious camembert cheese. Jérome simply asked me whether I would be interested to come and work some time in his laboratory. I was amazed, started laughing, explained him my toughts, and.. that was it. We agreed that I would come to Orsay and work for my Ph.D. there but that I would defend it in Belgrade. When my schollarship ended, I went back to Belgrade, completed and defended my M.Sc., prepared all the necessary administrative formalities, and in autumn of 1989., I was in Paris again, with a schollarship for 12 months.

Jérome's group was (and still is) one of the leading centers of studies of the organic metals, in particular of the Bechgaard salts. The name of these materials is derived from the fact that they were synthetized by K. Bechgaard and his group in Denmark [1], and their generic chemical formula is $(TMTSF)_2X$. The symbol $(TMTSF)_2$ denotes a complicated molecule of the name of di-tetra-meta-tetra-selena-fulvalene, which is the basic ingredient of all the Bechgaard salts, and X denotes an anion such as $PF_6, FSO_3, NO_3....$

When the Bechgaard salts were discovered, it was thought that they marked "the road" to be taken in the quest for high temperature superconductivity. This turned out to be wrong, so the motivation for their study (both experimental and theoretical) somewhat changed. The main motivation for their study nowadays is the idea that they are the simplest possible cases (and therefore easiest to understand) of correlated electron systems, which (it is thought worldwide) the high T_c materials are.

Time went quickly and I worked on the experiment aimed at measuring the electrical conductivity of the Bechgaard salt $(TMTSF)_2FSO_3$ [2]. When the experiment was almost complete, and we were starting to plan the continuation of it with the introduction of impurities, a critical part of the experimental setup exploded. No one was harmed, but waiting for the replacement I started trying to do some theoretical work on the conductivity of the Bechgaard salts. Jérome noted it and put me into contact with Heinz J.Schulz of the theory group. A couple of months before that, on advice of Jérome, I have asked for the prolongation of my schollarship for another year. This was at first rejected,but the Laboratory (and Jérome personally) used their "contacts" in the administration. After the summer holidays, when I got back to Paris, I found a hand-written note from Jérome saying "Dear Vladan, welcome for another year in Paris". I was proud and happy, went to the theory group to work with Schulz, and that is where my theoretical work on the organic conductors begun.

2   The Calculation of the Conductivity

2.1   The First Attempt

Heinz Schulz was a tall, often smiling, quiet, friendly man spending all the day in the laboratory, except on those days when he was giving lectures at the École Politechnique. He was from Hamburg, got his Ph.D.there and came to Orsay as a post-doc. He prepared another Ph.D. there,and was immediately offered a job. Sadly, he died of cancer in 1996., and only then I discovered that we were of nearly the same age. We agreed that after my experimental work in Jérome's group it was logical that I continue by trying to calculate the electrical conductivity of the Bechgaard salts.

Two immediate questions to which we had to find answers were the choice of the method of calculation,and the choice of the Hamiltonian (i.e.,a theoretical model of the material) which could be an appropriate description of the Bechgaard salts.

The logical choice of the theoretical model of these materials was the famous Hubbard model. It was proposed several decades before the start of my work in Orsay,and it contains the basic "ingredients" for the correct theoretical description of a solid:atoms, localized in the nodes of some form of a crystal lattice and electrons localized on them [3]. Although it is intuitively simple and acceptable, to this day the Hubbard model was solved only in the simplest (one-dimensional) case [4]. In one spatial dimension the Hamiltonian of the Hubbard model has the following form:

$$H = -t \sum_{i,\sigma=0}^{N} (c_{i+1,\sigma}^{+}c_{i,\sigma} + c_{i,\sigma}^{+}c_{i+1,\sigma}) + U \sum_{l=0} n_{l,\uparrow}n_{l,\downarrow} = H_0 + H_1 \qquad (1)$$

The first term is the kinetic energy term (H_0),and the second one is the interaction term (H_1). All the symbols in this equation have their standard meanings: t is the electron hopping energy,the terms in parenthesis in H_0 are the creation and anihillation operators for electrons on a lattice site i with spin σ. In the interaction term,U is the interaction energy of a pair of electrons having oposite spins within an atom on lattice site l.

A starting idea for the calculation of the conductivity was to try and apply the historic Kubo formula [5]. In modern transcription,such as [6], Kubo's formalism gives the following expression for the real part of the conductivity tensor:

$$\sigma'_{\mu\nu}(\omega) = \frac{\omega}{2V\hbar}(1 - exp^{-\beta\hbar\omega})I \tag{2}$$

and I denotes the integral:

$$I = \int_{-\infty}^{\infty} exp^{it\omega} \left\langle M_\mu(t)M_\nu(0)\right\rangle_0 \tag{3}$$

where M_μ is the time dependent dipole moment operator. In the Heisenberg picture,this time dependence is expressed as

$$M_\mu(t) = exp^{itH_0/\hbar}M_\mu exp^{-itH_0/\hbar} \tag{4}$$

To complicate things further, the calculation of the correlation function in the expression for I demands the knowledge of the partition function and the density matrix of the system. Obviously,the Kubo formula although being exact is too abstract for performing a realistic calculation.

2.2 The Memory Function Approach

After we jointly concluded that trying to apply Kubo's formula was actually leding to an *impasse* Schulz suggested that I try applying the so called "memory function" method. Of course I accepted,and had an immediate feeling that I was entering a sort of scientific "terra incognita".

The memory function method has its roots in work in pure rigorous statistical mechanics around the middle of the last century ([7],[8] and earlier work cited there). A theory of many-particle systems was developed in these papers with the aim of describing transport properties,collective motion and Brownian motion from a statistical-mechanical point of view. This formulation,taken from the abstract of [7], does not sound complicated. However,retracing in detail the calculation performed by Mori is a task of considerable mathematical complexity. The main point of using the memory function in any calculation of transport properties is that it gives the possibility of evading the problem of formulating and solving the transport equations. Instead of dealing with the transport equations, this method gives the possibility of expressing the conductivity in terms of a regular "memory function".

Papers [7] and [8] are,in some sense,comparable to [5]:they are exact in principle,but applicable to real calculations with extreme difficulty. However, some time after the publications of Mori, a technique was worked out which gave the possibility of applying the memory function method to real calculations [9],[10].

The equations needed for the calculation of the conductivity within the memory function method are [10]

$$\chi_{AB}(z) = << A;B >> = -i \int_0^\infty exp^{(izt)} < [A(t),B(0)] > dt \tag{5}$$

and

$$\sigma(\omega) = i\left(\frac{\omega_P^2}{4z\pi}\right)\left[1 - \frac{\chi(z)}{\chi_0}\right] \tag{6}$$

In these equations $\omega_P^2 = 4\pi n_e e^2 / m_e$ is the square of the plasma frequency, n_e, e and m_e are the electron number density,charge and mass,while χ_0 denotes the ratio n_e/m_e,which is the zero frequency limit of the dynamical susceptibility $\chi(\omega)$.

The integral in eq.(5) is a general definition of the linear response of an operator A to a perturbing operator B. It is an analytic function for all non-real frequencies z [10]. The symbol $A(t)$ denotes the Heisenberg representation of the operator A. Eq.(5) can "evolve" from a general definition of a response function into the definition of a current-current correlation function by introducing $A = B = [j,H]$. The current operator is denoted by j and H is the Hamiltonian of the system under consideration.

It was decided,in line with [11], to use the Hamiltonian of the Hubbard model in the calculation. The Hamiltonian of this model [3] in one dimension is given in eq.(1). The current operator has the form

$$j = -it\sum_{l,\sigma}(c_{l,\sigma}^+ c_{l+1,\sigma} - c_{l+1,\sigma}^+ c_{l,\sigma}) \tag{7}$$

The final step which needs to be made before the actual calculation of the conductivity of the Bechgaard salts is the determination of the chemical potential of the electron gas on a 1D lattice. It may not be obvious at first why is the chemical potential so important for this work. The usual research practice is that pure Bechgaard salts are described as systems with $\mu = 0$. Any possible doping (and experiments with doped specimen are very interesting) is theoretically described by deviating μ from 0. The chemical potential is a function of the band-filling n, the inverse temperature β and the electron hopping energy t. A determination of an analytical expression for this function demands a choice of a theoretical model for the description of the electrons in the Bechgaard salts: Fermi or Luttinger liquid?

2.2.1 Fermi or Luttinger Liquid?

The basic ideas of what is now (for several decades already) called the Fermi-liquid theory are present in physics for a long time - it can be safely stated since the time of Sommerfeld. He showed that various experimental data on metals (for example the low temperature behaviour of their specific heat or the electrical conductivity) can be understood by assuming that the electrons in a metal behave like a gas of non-interacting Fermions. Shortly after Sommerfeld, came the results of Pauli,Bloch and Wigner. They showed that the paramagnetic susceptibility of the non-interacting electrons is temperature independent, and that the interaction energies of the electrons at metallic densities are comparable to their kinetic energy. The "final touch" was given by Landau [12],[13]. In these works Landau introduced a new way of thinking about interacting systems, which became crucial for the developement of condensed matter physics. For example, the notions of quasiparticles and elementary excitations,were introduced in these papers. Methodologically, they were extremely important, because in them Landau introduced the idea of "asking useful questions about the low energy excitations of the system based on concepts of symmetry,without worrying about the myriad of unnecessary details" [14].

When does the Fermi liquid theory break down? Physically, the answer is logical: it happens whenever one of the measurable quantities which it aims to calculate diverges [14]. It has been shown in field theory that the Fermi liquid concept breaks down in one dimension. Technically speaking, some vertices which are in the Fermi liquid theory assumed finite, diverge in one dimension because of the Peierls effect (for example [15]). It is also known that the excitations in 1D are not quasi particles, but collective charge and spin degrees of freedom, each of which propagates with different velocity [16].

Optical and photoemission experiments on the Bechgaard salts show deviations from the Fermi liquid behaviour [17]. However,the energy scales on which the deviations occur in these two groups of experimental data are different. Interesting conclusions have been reached from c-axis conductivity measurements of the Bechgaard salts [18]. It was shown there that the 1D Luttinger liquid description of the Bechgaard salts breaks down below a pressure dependent value of the temperature,and that the Fermi liquid description is restored at low temperatures of the order of 10 K.

This subsection is certainly not a complete review of the problem "Fermi vs.Luttinger". It's aim is to try to justify to the reader the choice which was made in this work: to use the Fermi liquid theory. The logical question is "Why?".

There are several motives for this decision. First and simplest - chronologically,this calculation of the conductivity of the Bechgaard salts was completed in 1997., and started while the present author still was in Orsay. Second, at the time when this calculation was being performed many experimental data which are now avaliable simply did not exist. This implied that it is wiser to perform the calculation within the Fermi liquid theory as a very well known theoretical framework. Experimentally,it was known that the Bechgaard salts are not strictly 1D but quasi 1D - the conductivity along the so called c-axis is not exactly equal to zero. There is also the *a posteriori* justification: results for the temperature dependence of the resistivity are qualitatively similar to experimental data.

3   The Calculation-The Practical Part

3.1   The Chemical Potential

It is known since [4] that the chemical potential of the electron gas is zero for the single-band half-filled Hubbard model. The band filling is defined as the rato of the number of electrons present per lattice site n_e to the maximal possible number of electrons per site N. This value is obviously equal to 2 since two electrons with different spins can be present on the same lattice site.

$$n = \frac{n_e}{N} \tag{8}$$

Theoretically speaking, the value of n_e is a parameter which can be varied at will. From the experimental viewpoint, this value can be changed by doping the specimen with electron donors or acceptors. As the chemical potential enters the expression for the Fermi function, the first practical step in calculating the electrical conductivity had to be the determination of the chemical potential of the electron gas on a 1D lattice [19].

The starting point of this calculation is the following equation:

$$n = \frac{1}{\pi \hbar} \int \frac{1}{1 + exp\beta(\varepsilon_k - \mu)} dp \tag{9}$$

where all the symbols have their usual meaning, $p = \hbar k$ and n is the number of electrons per site. The integration is performed within the limits $\pm \pi \hbar / s$ and s is the lattice constant. In the 1D Hubbard model $\varepsilon_k = -2t \cos(ks)$, which leads to

$$dp = \hbar dk = \frac{\hbar}{s}(4t^2 - \varepsilon^2)^{-1/2} d\varepsilon \tag{10}$$

Inserting eq.(10) into eq.(9) gives

$$n = \frac{1}{\pi s} \int_{-2t}^{2t} \frac{1}{1 + exp\beta(\varepsilon_k - \mu)} (4t^2 - \varepsilon^2)^{-1/2} d\varepsilon \tag{11}$$

The initial problem of determining the function $\mu = \mu(\beta, t, n)$ has thus been reduced to the problem of solving the integral equation eq.(11). Instead of proceeding with the full rigour of the theory of integral equations, one can tackle the problem by using a suitable power series representation of the Fermi function.

The Fermi function can be represented as

$$\frac{1}{1 + exp\beta(\varepsilon_k - \mu)} = \Theta(\mu - \varepsilon) - \sum_{k=0}^{\infty} A_{2k+1} \beta^{-2(k+1)} \delta^{2k+1}(\varepsilon - \mu) \tag{12}$$

where the symbol A_{2k+1} denotes

$$A_{2k+1} = \frac{2(-1)^{k+2} \pi^{2k+2} (2^{2k+1} - 1)}{(2k+2)!} B_{2k+2} \tag{13}$$

Θ is the step function, $\delta^{2k+1}(\varepsilon - \mu)$ are the derivatives of the δ function and B_{2k+2} are the Bernoulli numbers.

Inserting eq.(12) into eq.(11) gives

$$n = \frac{1}{\pi s} \int_{-2t}^{2t} \left[\Theta(\mu - \varepsilon) - \sum_{k=0}^{\infty} A_{2k+1} \beta^{-2(k+1)} \delta^{2k+1}(\varepsilon - \mu) \right] (4t^2 - \varepsilon^2)^{-1/2} d\varepsilon \tag{14}$$

Eq.(14) is integrable by using the following relation, valid for any function $f(x)$ and its n-th order derivative:

$$\int \delta^n(x - x_0) f(x) dx = (-1)^n f^{(n)}(x_0) \tag{15}$$

Inserting eq.(15) in eq.(14) and limiting the summation to terms with $k \leq 2$ leads to an equation of second degree in μ which can be solved to give the following result for the chemical potential of the electron gas on a 1D lattice:

$$\mu = \frac{(\beta t)^6 (ns-1)\,|t|}{1.1029 + .1694(\beta t)^2 + .0654(\beta t)^4} \tag{16}$$

Obviously, for $\lim_{n,s \to 1}\mu = 0$, which means that this result has as its limiting value the well known result of Lieb and Wu [4].

The reader may at this point be tempted to ask for an explanation of the limitation to terms having $k \leq 2$ in the calculation leading to eq.(16).The explanation is that the idea being to obtain an analytical expression for the function $\mu(\beta, t, n)$ going to terms with $k \geq 2$ leads to expression which are too complicated for any applications in further work.

3.2   The Electrical Conductivity

With the chemical potential being known,it finally became possible to tackle the problem of the electrical conductivity of the Bechgaard salts [20]. The Hubbard hamiltonian and the current operator are given by eqs.(1) and (7).The functions $\chi(z)$ and $\sigma(z)$ are both expressible in complex form as $\sigma(z) = \sigma_R(z) + i\sigma_I(z)$ and $\chi(z) = \chi_R(z) + i\chi_I(z)$. If one further assumes that the frequency is a complex function,with $z = z_1 + iz_2$ and that $z_2 = \alpha z_1$ with $\alpha \succ 0$,it follows from eq.(6) that

$$\sigma_R + i\sigma_I = i\frac{\omega_P^2}{4z_1\pi(1+i\alpha)}\left[1 - \frac{\chi_R(z) + i\chi_I(z)}{\chi_0}\right] \tag{17}$$

This expression can be separated into the real and immaginary part. Taking the special case $\alpha = 0$ (justificed by the fact that the frequency measured in experiments is real) gives:

$$\sigma_R = \frac{\omega_P^2 \chi_I}{4z_1\pi\chi_0} \tag{18}$$

and

$$\sigma_I = \frac{\omega_P^2}{4\pi z_1}\left(1 - \frac{\chi_R}{\chi_0}\right) \tag{19}$$

The ensuing steps in the calculation of the electrical conductivity are in principle straightforward. The expression *"in principle"* is here amply justified, because inserting all the "sub-results" into the expression for χ_{AB} leads to an almost intractable expression.

The "practical" calculation (practical is here employed in the sense leading to the function σ_R) starts by determining the current-current correlation function.This can be obtained by inserting $A = B = [j, H]$ into eq.(5). Evaluating this commutator is facilitated to some extent by the decomposition of the the Hamiltonian of the Hubbard model indicated in eq.(1). Relatively easily it can be shown that $[j, H_0] = 0$ and that

$$A = [j,H] = [j,H_1] = -itU \sum_{l,\sigma} \left((c^+_{l,\sigma}c_{l+1,\sigma} - c^+_{l+1,\sigma}c_{l,\sigma}) (\delta_{l+1,j} - \delta_{l,j}) n_{j,-\sigma} \right) \quad (20)$$

All the symbols have their standard meanings. Transition to k space can be performed by relations of the following form [21]

$$c^+_{l,\sigma} = N^{-1/2} \sum_{k1} \exp(ik_1 ls) c^+_{k_1,\sigma} \quad (21)$$

In this expression N is the number of lattice sites, s is th elattice constant and $L = Ns$ is the length of the specimen. The temporal evolution of various operators can be introduced in the calculations by relation of the form $c_k(t) = \exp -i\varepsilon(k)t c_k$ where (for the one dimensional Hubbard model) $\varepsilon(k) = -2t\cos(ks)$. Inserting all this into the expression for χ_{AB} gives the following expression for the susceptibility $\chi(z)$

$$\chi(z) = \sum_{p,g,k,q} (32i(1/[(1+\exp(\beta(-\mu - 2t\cos(g))))$$
$$(1+\exp(\beta(-\mu - 2t\cos(k))))] - 1/(1+\exp(\beta(-\mu - 2t\cos(p))))$$
$$1/(1+\exp(\beta(-\mu - 2t\cos(q)))))](Ut)^2(\alpha z_1 + i(z_1 + 2t(\cos(q) + \cos(p)$$
$$-\cos(g) - \cos(k))))(\cos(p+g)/2)(\cos((q+k)/2)[\cosh(g-p)-1]/$$
$$(N^4((\alpha z_1)^2 + (z_1 + 2t(\cos(q) + \cos(p) - \cos(g) - \cos(k))))^2)) \quad (22)$$

In order to simplify the calculation at least to some extent, the summation was limited to the first Brillouin zone, and the lattice constant s was set as $s = 1$. This obviously implies that $L = N$. The sum in the equation above were calculated under the condition $\alpha \neq 0$ because this condition is built in in the definition of $\chi(z)$. Once the sums were calculated, taking into account that in reality the frequency is a real quantity, the limit $\alpha \to 0$ was imposed. After performing all the summations and taking $\alpha \to 0$ the following approximation for the dynamical susceptibility was obtained.

$$\chi \cong (32i(-1+\cosh(1))[1+\exp(\beta(-\mu + 2t\cos(1-\pi)))]^{-2}$$
$$(1+\exp(\beta(-\mu - 2t)))^{-2}](Ut/N^2)^2 \cos^2((1-2\pi)/2)$$
$$(z_2 + i(z_1 + 2t(-2 - 2\cos(1-\pi))))/(z_2^2 + (z_1 + 2t(-2 - 2\cos(1-\pi)))^2$$
$$+ << 2267 >> \quad (23)$$

The number $<< 2267 >>$ in the preceeding equation denotes the number of omitted terms. Obviously, such a long equation is inapplicable and has to be truncated after a certain number of terms. Taking the first 32 terms of eq.(23), multiplying out the products and powers, expressing the result as a sum, gives the real part of the dynamical susceptibility as

$$\chi_R(z) \cong [128U^2t^3 \cos^2((1-2\pi)/2)]/[1+\exp(\beta(-\mu - 2t)))^2 N^4(z_2^2 + $$
$$+(z_1 + 2bt)^2)] + << 527 >> \quad (24)$$

and $b = -4(1+cos(1-\pi))$.

Taking the first 20 terms of eq.(24) and imposing the condition $\lim z_2 \to 0$ it follows that the real component of the function χ_R has the form

$$\chi_R(\omega) = \sum_i \frac{K_i}{(\omega+bt)^2} + \sum_j \frac{L_j\omega}{(\omega+bt)^2} \qquad (25)$$

The functions K_i and L_j can be read-off from eq.(24) developed to a given number of terms. Using this result, the immaginary part of the dynamical susceptibility (χ_I) is given by [6]

$$\chi_I(\omega) = -2\frac{\omega_0}{\pi}P\int \frac{\chi_R(\omega)}{(\omega^2-\omega_0)^2} \qquad (26)$$

Inserting eq.(25) into eq.(26) and imposing the constraint $i,j \le 4$ in eq.(25) leads to the following expression for χ_I

$$\chi_I = (2bt/\pi)(Ut/N^2)^2[\omega_0/(\omega_0+2bt)(\omega_0^2-(2bt)^2)][4.53316(1+$$
$$\exp(\beta(-\mu-2t)))^{-2}+24.6448(1+\exp(\beta(-\mu+2tcos(1-\pi)))))^{-2}]+$$
$$(2/\pi)[\omega_0/(\omega_0^2-(2bt)^2)](Ut/N^2)^2 \times [42.49916(1+\exp(\beta(-\mu-2t)))^{-2}+$$
$$78.2557(1+\exp(\beta(-\mu+2t\cos(1-\pi))))^{-2}]$$

$$(27)$$

The final expression for the electrical conductivity follows by inserting this result into eq.(18). After some algebra, one gets the final result for the electrical conductivity of Q1D organic metals

$$\sigma_R(\omega_0) = (1/2\chi_0)(\omega_P^2/\pi)[\omega_0^2-(bt)^2]^{-1}(Ut/N^2)^2 \times S \qquad (28)$$

and the symbol S denotes the following function

$$S = 42.49916 \times (1+\exp(\beta(-\mu-2t)))^{-2}+78.2557 \times$$
$$(1+\exp(\beta(-\mu+2t\cos(1+\pi))))^{-2}+(bt/(\omega_0+bt)) \times$$
$$(4.53316 \times (1+\exp(\beta(-\mu-2t)))^{-2}+$$
$$24.6448(1+\exp(\beta(-\mu+2t\cos(1+\pi)))))^{-2})$$

$$(29)$$

3.3 Discussion

Equations (28) and (29) are the final result for th electrical conductivity of the Q1D organic metals obtained within the memory function method. From the purely mathematical point of view they are "moderately complicated", the main "'complicating factor" being the length of the two expressions. However, for applications in physics, the relevant question is whether or not the results of these two expressions agree with experiments.

The main externally controllable parameters in experiments on organic conductors are the temperature, the doping and the frequency, and in the remainder of this chapter, the comparisons of the results of eqs.(28) and (29) will be performed with respect to these parameters. Model parameters which are constant in these two expressions were chosen as follows: $N = 150, U = 4t, \omega_P = 3U, \chi_0 = 1/3$ and $\omega_0 \geq 0.6U$. The lower limit for ω_0 was determined by imposing the condition that $\sigma_R \geq 0$, while all the other values were chosen in analogy with high temperature superconductors. The influence of doping (and its changes) on the conductivity can be studied through variations of the chemical potential which depends (among other parameters) on the band filling. A half-filled band ($n = 1$) describes a chemically pure specimen. Positive deviations of the filling from the value $n = 1$ experimentally correspond to doping with electron donors. Negative deviations describe the doping of a specimen by electron acceptors. The first test of the result obtained for the electrical conductivity was its application to the case of a 1D half-filled Hubbard model. Namely, it is known from rigorous theory that its conductivity is zero (for example [22]). Inserting $n = 1$ into eq.(29) and developing it in t as a small parameter, it follows that the conductivity is approximately given by

$$\sigma_R(\omega) \cong 10^{-7}(\frac{\omega_P}{\omega})^2 t^4 (4.56 + .3\beta t) \tag{30}$$

Clearly, the numerical value of this expression is close to zero for physically acceptable values of the input parameters, of course excluding the case $\omega = 0$. Looking from the experimental side, this can be interpreted as implying that weakly conducting phases of the Bechgaard salts can be described by a 1D Hubbard model with a small deviation of the band filling from the value of $1/2$.

The experimental parameter which can be most easily controlled in experiments on organic conductors is the temperature of the specimen. The general conclusion of all such experiments is that the temperature dependence of the conductvity of the Bechgaard salts is extremely complex, and that it can not be explained by conventional theory of conductivity of metals.

In the calculations discussed in this chapter no attempt was made to reproduce the experimental conductivity of any particular salt. However, the idea was to determine whether or not the one band 1D Hubbard model can reproduce semiquantitatively the experimental data. Examples of real experimental data and theoretical calculations of the conductivity are avaliable in [17],[20] and many other publications.

Equations (28) and (29) can be re-written as follows

$$\sigma_R(\omega_0) = A \left[\omega_0^2 - (bt)^2 \right]^{-1} \left[Q + btZ(\omega_0 + bt)^{-1} \right] \tag{31}$$

where the sumbols A,Q and Z denote various frequency independent functions which occur in the expression for the conductivity. Developing this result in ω_0 up to second order terms, it follows that

$$\sigma_R(\omega_0) \cong -A(Q+Z)(bt)^{-2} + AZ(bt)^{-3}(\omega_0) - A(2Z-Q)(bt)^{-4}(\omega_0)^2 + \dots \quad (32)$$

Fitting this equation to measured frequency profiles of the conductivity, one could determine the functions A,Z,Q and t. An approximate value of the static limit of the function $\sigma_R(\omega_0)$ follows from eq.(32) as

$$\sigma_R(\omega_0 = 0) \cong -A(Q+Z)(bt)^{-2} \quad (33)$$

Fitting these two expressions to experimental data, it would be possible to determine the values of the functions A,Z,Q and the hopping integral t. Fixing all the parameters and varying the band filling would give the possibility of investigating the effects of doping on the conductivity. A preliminary investigation of this sort has been performed in [20].

3.4 The Equation of State and Specific Heat

It is hoped that the reader has by now gained an idea about the complexity of theoretical studies of the Bechgaard salts. The aim of this sub-section is to outline a determination of the specific heat per particle under constant volume of a degenerate electron gas on a 1D lattice. Such a calculation may seem as an exercise in statistical physics and mathematics. It fact, in recent years measurements of specific heat and thermal transport properties of high T_c superconductors and organic conductors have become a useful tool in studies of these systems. The literature in the field is steadily growing, and [23],[24],[25] are some useful examples.

It will be assumed in the following that the number of particles N in the system is not conserved. Mathematically speaking, the starting point of the calculations is the equation

$$dG = -SdT + VdP + \mu dN \quad (34)$$

where $G = \mu N$ and all the symbols have their standard meanings. Inserting the definition of the thermodynamic potential G into eq.(34) and differentiating with respect to T, one obtains

$$V(1 + \frac{1}{V}\frac{\partial V}{\partial T})\frac{\partial P}{\partial T} = S + \frac{\partial S}{\partial T} + \frac{\partial N}{\partial T}\frac{\partial \mu}{\partial T} + N\frac{\partial \mu}{\partial T} \quad (35)$$

In the last expression the obvious transformation $dP = \frac{\partial P}{\partial T}dT$ (and similar relations for other variables) was applied. Eq.(35) is expressed in terms of the bulk parameters of the system. In order to re-write it in terms of the local variables, a change of variables $N \to nV$ and $S \to nsV$ is necessary. After some algebra, it can be shown that

$$V(1 + \frac{1}{V}\frac{\partial V}{\partial T})\frac{\partial P}{\partial T} = ns(V + \frac{\partial V}{\partial T}) + V\frac{\partial n}{\partial T}(s + \frac{\partial \mu}{\partial T}) +$$
$$nV\frac{\partial}{\partial T}(s + \mu) + n\frac{\partial \mu}{\partial T}\frac{\partial V}{\partial T} \quad (36)$$

We have here derived the differential equation of state (EOS) of any material. It is at first sight complicated, but in applications to the Bechgaard salts it can be considerably simplified. Experiments on these materials are almost always performed under fixed volume conditions [26]. This implies that all terms in eq.(36) containing volume derivatives can be disregarded. The final form of the EOS of Q1D organic metals thus emerges as

$$\frac{\partial P}{\partial T} = (n + \frac{\partial n}{\partial T})(s + \frac{\partial \mu}{\partial T}) + n\frac{\partial s}{\partial T} \tag{37}$$

The specific heat per particle is given by [27]

$$c_V = \frac{T}{n}(\frac{\partial^2 P}{\partial T^2})_V - T(\frac{\partial^2 \mu}{\partial T^2})_V \tag{38}$$

Applying this definition and performing all the necessary algebra, one finaly gets the expression for the specific heat per particle

$$c_V = \frac{T}{n}(s + \frac{\partial \mu}{\partial T})(\frac{\partial n}{\partial T} + \frac{\partial^2 n}{\partial T^2}) + \frac{T}{n}\frac{\partial n}{\partial T}(2\frac{\partial s}{\partial T} + \frac{\partial^2 \mu}{\partial T^2}) + $$
$$T(\frac{\partial s}{\partial T} + \frac{\partial s^2}{\partial T^2}) \tag{39}$$

In order to apply eq.(39) to the degenerate electron gas on a 1D lattice, it is necessary to introduce into it appropriate expressions for the chemical potential, number density and entropy per particle. The chemical potential is given by eq.(16). As a first approximation, known results for the entropy per particle and the number density can be used [28]. The entropy per particle is given by

$$s = \frac{Q}{n}\frac{\partial F_{3/2}(\mu/T)}{\partial T} \tag{40}$$

The symbol $F_{3/2}(\mu/T)$ denotes a special case of a Fermi-Dirac integral [28], n is the electron number density and Q is a combination of known constants (such as the electron mass and the Planck constant). Finally, the number density of a degenerate electron gas at low temperature can be expressed as [28]

$$n \cong A \times T^{15/2}\left[1 + B \times T^{3/2} + ...\right] \tag{41}$$

where the symbols A and B denote combinations of known constants. Inser- ting eqs.(16),(40) and (41) into eq.(39) gives the final result for the specific heat of the electron gas on a 1D lattice. After all the necessary algebra, it would turn out to be highly non-linear, in line with the results of various experiments (such as [24],[29]and later work).

4   Conclusion

The aim of this chapter was to review some results of theoretical studies of a class of organic conductors called the Bechgaard salts. From the start it was prepared as a "personal review",which simply means that it is to a large extent based on previous research results of the author. At the end of such a review,one can reflect for a moment on the possible future developement of research on the Bechgaard salts, and the organic conductors in general. Like Lord Kelvin in the *XIX* century, one may be tempted to think that after more than 20 years since their discovery, most things about the Bechgaard salts are well known and only "small clouds" remain to be clarified. It seems (at least to the present author) that this would be totally wrong,and many interesting problems are waiting to be solved. A few of them are mentioned in the follow up.

The general question can be formulated as "Why should anybody on this planet study the organic conductors (and the Bechgaard salts as a particular family of the organic conductors)?" The answer is "simple" : the Bechgaard salts are a Q1D example of strongly correlated systems,and therefore should be easier to understand than their 3D analogs,such as the high T_c superconductors. However,in spite of all the efforts, the physical mechanism of high T_c superconductivity has not yet been discovered.

A basic problem is, of course, the applicability (or inapplicability) of the Fermi liquid model to these materials. Indications from various experiments (for example [17]) seem to be that the Fermi liquid is not a good description of the organic conductors. To make things more interesting, there are also indications (such as [20] or [18]) that reasonable agreement between calculations performed within the Fermi liquid model and experiments can be achieved in some cases. A definite solution of this dilemma is certainly one of the problems waiting to be solved as soon as possible.

The calculations reported in the present chapter were performed within the "original version" of the memory function method. The method itself has been considerably reformulated and modernized in the mean time, bringing it more "in line" with contemporary field theory. For a recent application of a modernized version of the memory function method see, for example, [30]. Another method which may be useful for calculations like those described in this chapter(but also for the determination of the phase diagram) has been developed in [31].

The determination of the phase diagram of the Bechgaard salts is an interesting problem on its own. The temperature dependence of their conductivity can vary in some regions of the $P - T$ plane; accordingly, they can be insulators, superconductors but also normal conductors or semiconductors. The general form of the phase diagram of the Bechgaard salts is known for several decades. However,much more complicated is the possibility of determining the parameters (such as t) of a Bechgaard salt from the analysis of the phase boundaries on a $P - T$ diagram.For a recent example of such a work see [32].

Interesting considerations have recently been made concerning the dimensionality of the Bechgaard salts. Namely,they are usually considered as one dimensional,or quasi one-dimensional. However,recent angular magnetoresistance osicllation (AMR) experiments [33] on $(TMTSF)_2FSO_3$ have shown that a cylindrical Fermi surface can be formed for this material. Does this perhaps mean that the Bechgaard salts should be considered as Q2D materials is (to the knowledge of the present author) a completely open question.

This list of selected problems concerning the Bechgaard salts could be continued, but it is hoped that the examples present illustrate sufficiently well that work on these materials is an interesting field of condensed matter physics.

Acknowledgement

This review was prepared within the project 1231 financed by the Ministry of Science and Protection of the Environment of Serbia.

References

[1] Bechgaard,K.,Jacobsen,C.S.,Mortensen,K.et al.: *Solid State Comm.*, **33**,1119 (1980)

[2] Auban,P.,Čelebonović,V.,Tomić,S., Jérome,D. and Bechgaard,K.: *Synth.Metals*,**41**-43,2281 (1991).

[3] Hubbard,J.:*Proc.Roy.Soc.*,**A276**,238 (1963)

[4] Lieb,E.H. and Wu,F.Y.: *Phys.Rev.Lett.*,**20**,1445 (1968).

[5] Kubo,R.: *J.Phys.Soc.Japan*,**12**,570 (1957)

[6] Wallis,R.F. and Balkanski,M.: *Many-body aspects of solid state spectroscopy*,North Holland,Amsterdam (1986).

[7] Mori,H.:*Progr.Theor.Phys.*,**33**,423 (1965).

[8] Mori,H.:*Progr.Theor.Phys.*,**34**,399.(1965).

[9] Götze,W. and Wölfle,P.: 1971,*J.Low Temp.Phys.*,**5**,575 (1971).

[10] Götze,W. and Wölfle,P.:*Phys.Rev.*,**B6**,1226 (1972)

[11] Anderson,P.W.:*Science*,**235**,1196 (1987).

[12] Landau,L.D.: *Sov.Phys.-JETP*,**3**,920 (1957a).

[13] Landau,L.D.: *Sov.Phys.-JETP*,**5**,101 (1957b).

[14] Varma,C.M.,Nussinov,Z.and van Saarloos,W.:*Phys.Rep.*,**361**,267(2002).

[15] Voit,J.:preprint *cond-mat/0005114* (2000).

[16] Voit,J.: *J.Phys.:Condens.Matter.*,**5**,8305 (1993).

[17] Vescoli,V.,Zwick,F.,Henderson,W. et.al.: *Eur.Phys.J.*,**B13**,503 (2000).

[18] Moser,J.,Gabay,M.,Auban-Senzier,P. et.al.: *Eur.Phys.J.*,**B1**,39 (1998).

[19] Čelebonović,V.: *Phys.Low-Dim.Struct.*,**11/12**,25 (1996).

[20] Čelebonović,V.: *Phys.Low-Dim.Struct.*,**3/4**,65 (1997).

[21] Emery,V.J.:1979,in *"Highly Conducting One Dimensional Solids"* (ed.by J.T.Devreese,R.P.Erward and V.E.van Doren),p.247, Plenum Press,New York 1979.

[22] Schulz,H.J.: preprint *cond-mat/9503150* (1995).

[23] Yu,R.C.,Salamon,M.B.,Lu,J.P. and Lee,W.C.: *Phys.Rev.Lett.*,**69**,1431 (1992).

[24] Lasjaunias,J.C.,Biljaković,K.,Yang.H. and Monceau,P.:*Synth.Metals*, **103**,2130 (1999).

[25] Behnia,K.,Belin,S.,Aubin,H. et al.: *J.Low Temp.Phys.*,**117** 1089 (1999).

[26] Ishiguro,T.and Yamaji,K.:*Organic Conductors*,Springer Verlag,Berlin (1990).

[27] Stanley,H.E.: *Introduction to Phase Transitions and Critical Phenomena*,Clarendon Press,Oxford (1971).

[28] Čelebonović,V.:preprint *astro-ph 9802279* (1998).

[29] Belin,S.and Behnia,K.: *Phys.Rev.Lett.*,**79**,2125 (1997).

[30] Kupčić,I.: *Physica* **B244**,27 (2004).

[31] Scalapino,D.J.,White,S.R. and Zhang,S.: *Phys.Rev.*,**B47**,7995 (1993).

[32] Čelebonović,V.: preprint *cond-mat/0402127*

[33] Kang,W.,Jo,Y.J.,Kang,H et al.: *Physica*,**C388**-389,585 (2003).

INDEX

A

absorption, viii, 81, 91, 92, 93, 94, 95, 96, 97, 98, 120, 123, 128, 140, 143
acetone, 224, 225
acrylic acid, 205
additives, 192, 193, 194, 205, 211, 215
adhesion, 227
adsorption, xii, 194, 195, 199, 209, 219, 232, 233, 239, 240, 241
aggregates, xi, 158, 188
aging, 196, 197, 198, 199, 202, 208
alcohol, 225, 229, 230, 237
algorithm, 46, 102
alloys, viii, ix, x, 81, 82, 83, 84, 85, 86, 87, 88, 89, 90, 91, 92, 93, 94, 95, 96, 97, 98, 101, 111, 114, 115
aluminum, 10
ambient air, 233
anisotropy, vii, xi, 1, 2, 3, 8, 13, 14, 16, 17, 19, 20, 21, 23, 26, 29, 30, 32, 33, 34, 39, 42, 44, 219, 221, 223
annealing, 150, 151, 152
argon, 84, 222, 223
asymmetry, 30, 32, 33, 34, 36
atmospheric pressure, 230, 231, 234
atomic force, 50, 64
atomic force microscope (AFM), 50

B

background noise, 194
Bechgaard salts, xii, 245, 246, 247, 249, 250, 252, 255, 256, 257, 258
bending, 5, 10, 16, 55, 59
biocompatibility, 82
Boltzmann constant, 130
boundary surface, xii, 219, 221, 235, 236, 237
breakdown, 186, 230, 233, 234

C

calibration, 194
calorimetry, vii
carbon, 197
carriers, 119, 121, 127, 130, 132, 134, 136, 138, 139, 140, 142, 143, 144, 227
cell, 111, 160, 161, 178, 179, 181, 187, 192, 194
ceramic, 82, 158, 178, 191, 192, 193, 195, 207, 208, 209, 211, 217, 220, 236
CH_3COOH, 194
charge trapping, 220
chemistry, vii
classification, 76
clusters, 239, 241
CO_2, 197, 201, 227
coating, 195, 211
cobalt, vii, 1, 17, 18
combined effect, 143
compatibility, 111
composites, vii
composition, 4, 82, 83, 93, 98, 195, 207, 215, 233
computer simulations, 188
condensation, 45
conductivity, 2, 5, 31, 42, 43, 176, 220, 222, 223, 237, 239, 245, 247, 248, 249, 250, 255, 256, 258
constant load, 11
cooling process, 103
copper, 6, 8
corona discharge, xi, xii, 219, 221, 223, 224, 227, 228, 229, 230, 231, 237, 238
correlation function, 248, 249, 252
corrosion, 82, 189
Coulomb interaction, 121, 128, 143
covering, 55, 56, 57, 184
crystal growth, vii

cubic system, x, xi, 157, 158, 159, 160, 169, 172, 176, 177, 178, 179, 181, 182, 185, 187
curing, 85

D

decay, xi, xii, 46, 117, 134, 219, 220, 221, 222, 223, 224, 225, 226, 227, 228, 229, 230, 231, 236, 237, 239, 241, 242
decay times, 134
decomposition, 252
defects, 16, 115, 149, 155
deformation, x, 157, 158, 160, 161, 178, 179, 188
degradation, 9
density, ix, xi, 4, 5, 7, 8, 9, 10, 12, 31, 45, 118, 119, 126, 127, 128, 129, 133, 143, 144, 158, 173, 178, 185, 191, 192, 195, 196, 205, 206, 207, 208, 213, 214, 215, 233, 239, 240, 248, 249, 257
density of states, 126, 127, 128, 133
density values, 205
desorption, xii, 219, 231, 232, 237
dielectric constant, 126, 132, 234, 235, 236
dielectrics, 220, 237, 243
diffuse reflectance, 194
diffusion, 42, 44, 192, 195, 200, 201
diffusion time, 42
discharges, 227
displacement, 11
distilled water, 223, 225, 229, 230, 237

E

efficiency, x, 192, 209
electric field, 5, 42, 220
electrical conductivity, xii, 241, 245, 247, 249, 251, 252, 254, 255
electrical resistance, 181, 183, 184, 185, 186
electrodes, xi, 195, 219, 221, 222, 224, 225, 229, 230, 231, 232, 233, 234, 236, 238
electron microscopy, vii
electrons, viii, ix, 3, 81, 82, 92, 95, 96, 97, 98, 129, 247, 249, 250, 251
elongation, 205
emission, 119, 125, 126, 127, 128, 130, 131, 132, 133, 134, 135, 136, 138, 140, 141, 142, 143, 149, 151
energy, vii, x, 1, 4, 22, 30, 31, 41, 44, 46, 92, 93, 103, 104, 110, 117, 118, 119, 121, 122, 123, 126, 127, 128, 130, 131, 132, 133, 134, 135, 136, 137, 138, 139, 140, 141, 142, 143, 192, 208, 229, 247, 249, 250
energy density, 22, 30, 41
engineering, vii, 3, 102, 205
entropy, 257

ethylene oxide, 197
evaporation, 149
excitation, 17, 43, 118, 119, 143, 144
extraction, viii, 49, 71, 74
extrapolation, 17

F

fabrication, x
fatigue, viii, 49, 51, 54, 55, 56, 57, 59, 60, 61, 62, 63, 64, 65, 66, 67, 69, 70, 75, 76, 77
Fermi level, 92
Fermi surface, 258
fermion, 122
ferromagnets, 2, 4, 16, 23
field theory, 250, 258
films, 181, 183, 184
flammability, 192
flocculation, 205
foils, 227
fractal analysis, viii, 49, 50, 51, 54, 55, 56, 58, 59, 62, 63, 65, 66, 67, 68, 76, 77
fractal dimension, viii, 49, 50, 51, 55, 56, 57, 58, 62, 63, 64, 65, 66, 67, 68, 69, 70, 71, 73, 74, 75, 76, 77

G

gas tungsten arc welding (GTAW), ix
gold, viii, ix, 81, 82, 90, 93, 94, 95, 96, 98, 195
grain boundaries, 76, 207
grains, 2, 3, 16, 30, 39, 40, 59, 63, 64, 66, 103, 104, 105, 108, 206
gravitational effect, ix
gravitational orientation, ix
gravity, ix

H

Hamiltonian, 138, 247, 249, 252
hardness, 4
heat, ix, x, 44, 59, 84, 185, 186, 187, 199, 200, 201, 202, 215, 249, 256, 257
heating rate, 149, 151, 229
Heisenberg picture, 248
homogeneity, 192
hydrogen, 121, 130
hydrothermal synthesis, 200
hydroxide, 194
hydroxyl, 200, 205, 206
hysteresis loops, 20, 42, 44

I

impurities, 148, 149, 155, 156, 184, 193, 247
inflationary sphere model, x, 157, 160
initial state, 122, 143
insulators, 258
interface, 59, 61
ionization, 221
ionizing radiation, 147, 148
ions, xi, 148, 149, 155, 191, 192, 194, 197, 199, 200, 205, 208, 209, 215, 221, 227, 231, 232
IR spectra, 200
irradiation, 149, 150, 155
isotherms, 195, 209, 210

K

kinetic parameters, 155
kinetics, 148, 154, 156, 228, 239, 241
KOH, 194

L

lamination, 196, 206
laser beam welding (LBW), ix
linear dependence, 32, 41
liquid phase, 184
low temperatures, xi, 119, 144, 219, 224, 226, 250
luminescence, 155

M

magnetic fields, vii, 1, 2, 4, 6, 7, 8, 10, 12, 14, 15, 16, 17, 19, 21, 22, 23, 29, 30, 31, 35, 36, 38, 39, 40, 41, 42, 44, 45, 46
magnetic moment, 4, 46
magnetism, 7
mass, 4, 118, 120, 127, 128, 129, 130, 132, 133, 135, 140, 142, 144, 148, 158, 184, 185, 249, 257
mechanical properties, x, 50, 191, 194, 205, 206, 215
medicine, 172
melt, 148, 149, 151, 184, 185
melting, x, 148, 185
melting temperature, x
meson, 46
metallurgy, 158, 172
metals, vii, viii, x, 49, 50, 63, 66, 67, 68, 76, 81, 82, 84, 86, 89, 90, 91, 92, 95, 98, 115, 189, 237, 246, 249, 254, 255, 257
Mg, 148
microscope, viii, 2, 50, 59, 64, 71, 77, 195
microstructure, viii, ix, x, 50, 195, 196, 204, 206, 207, 212

microstructures, 50, 51, 63, 65, 66, 67, 68, 73, 76, 77, 114, 206
mobility, 121, 122, 133, 134, 144, 228
modeling, x, 2, 157, 158, 184, 185, 188
moisture, 241
molar ratios, 197
molecular weight, 209
momentum, 4, 120, 123, 124, 125, 128, 140
monomolecular, 232
morphology, ix, x, 65, 77, 170, 180, 193, 204, 205, 206, 209, 213, 215

N

neutron, 46
neutrons, 46
nickel, vii, 1, 17, 18, 183
nitrogen, 148, 229
nonequilibrium, 32, 127
nucleation, 32
nuclei, 239, 240, 241

O

optical properties, viii, 81, 82, 84, 95
organic solvents, 192
orientation, ix, 105, 110, 113
oxidation, 189
oxides, 184
oxygen, 228, 229

P

palladium, 82
particles, x, 61, 157, 158, 159, 160, 161, 169, 170, 171, 178, 181, 184, 185, 186, 188, 192, 193, 194, 250, 256
passivation, xi, 191, 192, 195, 211, 215
percolation, xii, 219, 221, 237, 239, 240, 241
percolation cluster, 239, 240
percolation lattice, 239
percolation theory, 239, 241
permeability, 8, 9, 42, 43, 44, 172
permittivity, 207, 208, 209, 213, 214, 215
PET, 227, 236
pH, xi, 191, 194, 196, 197, 198, 199, 205, 208, 210, 215
phase diagram, 258
phase transformation, ix, 4, 101, 103, 107, 108, 109, 110, 112, 113, 114, 115
phase transitions, 2
photoelectron spectroscopy, 229
photoemission, 250

photoluminescence (PL), 117, 118, 119, 127, 130

photon, 119, 120, 122, 123, 126, 131, 137, 138, 140, 143

photons, 95, 120, 122, 127

physical properties, 82

plasma, 249

plastic deformation, 107

plasticity, 102, 103, 104, 105, 107, 108

plasticizer, 193, 205

platinum, x, 148

PMMA, 197, 198, 209, 210, 211, 212, 213, 214

Poisson ratio, 108

polarity, 221, 228

polarization, 45, 120, 220, 228, 229

pollution, xi, 219, 221, 223, 224

polyether, 197, 205, 206

polymer, xi, xii, 219, 221, 225, 227, 228, 229, 230, 236, 237

polymers, vii, 192, 222

polymethylmethacrylate, 197

polypropylene (PP), 194, 225, 229, 230, 237

porosity, x, xi, 157, 158, 159, 160, 161, 164, 166, 167, 168, 169, 170, 172, 173, 174, 175, 176, 177, 178, 179, 180, 181, 183, 184, 185, 188, 191

properties, vii, viii, x, xi, 10, 67, 81, 82, 84, 98, 155, 191, 192, 193, 194, 195, 196, 205, 207, 208, 211, 215, 220, 227, 248, 256

protons, 200

PTFE, 149, 225, 227, 228, 236, 237

purification, 224

Q

quantum state, 119

R

radiation, 91, 147

radicals, 232

rate constants, 232

raw materials, 191

reduction, xi, 8, 23, 33, 191, 192

refractive index, 120

regression, 55, 56, 57, 240

reproduction, 144

research, vii, viii, x, 3, 8, 35, 46, 50, 71, 98, 188, 191, 200, 245, 246, 249, 258

resin, 85, 194, 197, 230

resistance, 82, 181, 190

reversal hysteresis curves, vii, 1

room temperature, 183, 194, 197, 199, 207, 225, 228, 237, 238, 242

roughness, viii, 50, 51, 58, 64, 68, 69, 70, 71, 73, 74, 75, 76, 77

rubber, 28

S

salts, xii, 245, 250, 258

scaling, 110

scanning tunneling microscope, 64

scattering, 46

self-similarity, viii, 49, 68

semiconductors, 118, 119, 121, 123, 126, 128, 131, 138, 140, 143, 220, 258

shape, vii, viii, ix, x, 1, 2, 5, 7, 8, 26, 31, 32, 44, 50, 51, 55, 58, 68, 69, 71, 73, 74, 76, 77, 87, 101, 103, 107, 111, 114, 115, 150, 157, 158, 169, 174, 195, 225, 240

shear rates, 203

silicon, 59, 60, 61, 62, 71, 72, 117, 144

silver, 195

smoothing, 175

solid phase, 173

solid state, 259

solubility, 197, 207

solution, xi, 9, 105, 159, 160, 191, 192, 194, 196, 197, 198, 215, 232, 234, 258

solvent, 192

Spain, 157, 190

species, 184, 192, 194, 195, 197, 200, 205, 212, 213, 227

specific surface, 200

spectroscopy, 95, 117, 259

spin, 118, 129, 142, 247, 250

steel, vii, ix, 1, 5, 8, 9, 10, 13, 14, 15, 16, 17, 18, 19, 22, 23, 24, 25, 26, 27, 28, 29, 31, 33, 37, 38, 40, 42, 43, 46, 50, 59, 60, 66, 67

stereo matching method, viii, 49, 50, 51, 52, 55, 59, 61, 64, 65, 68, 76, 77

strength, x, 7, 233

stress fields, vii, 1, 32, 33, 36, 41, 46

surface area, 193, 200

surface layer, 229

suspensions, 205

symmetry, 38, 148, 250

synthesis, 211

T

temperature, vii, x, 1, 2, 3, 10, 17, 22, 31, 32, 42, 44, 86, 102, 103, 110, 114, 117, 119, 126, 130, 134, 135, 136, 138, 141, 142, 143, 148, 149, 174, 186, 194, 200, 201, 202, 206, 220, 222, 224, 226, 227, 228, 229, 242, 246, 249, 250, 255, 257, 258

temperature annealing, 148
temperature dependence, 141, 142, 143, 227, 228, 229, 255, 258
temperature dependences, 229
tensile strength, 204, 205
thermal activation, 149, 156
thermal properties, xii, 245
thermal relaxation, 5, 121
thermal treatment, 174
thermodynamic equilibrium, 240
thermodynamics, 2, 3, 4, 32
thermograms, 228
thermoluminescence, 148, 150
thin film, 114, 184, 244
thin films, 184
titanium, 183
toxicity, 192
transformation, ix, 4, 101, 102, 103, 110, 111, 112, 113, 114, 115, 256
transition metal, 92
transport, xi, 186, 219, 220, 221, 226, 227, 239, 248, 256
transport processes, xi, 219, 221, 226
treatment, 84, 184, 199, 200, 202, 215

U

ultrasonic vibrations, 9
unified field theory, 31, 42

V

vacuum, 222, 229, 230, 231
vector magnetizations, vii, 1
viscosity, xi, 15, 32, 191, 192, 193, 203, 205, 211, 212, 213, 215

W

water, xi, 86, 191, 192, 193, 194, 195, 197, 198, 199, 200, 201, 209, 215, 216, 221, 237, 239, 240
wave vector, 126
wavelengths, viii, 81, 87, 88
wires, 5

X

x-ray diffraction, vii

Z

zirconia, 192, 194